PRAISE FOR

PIECES O

"Fernyhough is a gifted writer who can t
... The stories in *Pieces of Light* ... will

"As absorbing as it is thought provoking." —*Sunday Business Post* (Dublin)

"Weaving scientific research from psychology, neuroscience, and evolutionary biology, Fernyhough explains that our brains don't record experiences as cameras do; rather, we store key elements, then reconstruct the experiences when we need them, imbuing them with present-day feelings and the benefit of hindsight." —*Washington Post*

"In this lyrical exploration of our powers of recall, psychologist and novelist Charles Fernyhough argues that our memories are worth cherishing—even though some of what we think we remember is, in fact, fiction." —*New Scientist*, Books of the Year

"Remarkable storytelling skills.... Seamlessly intersperses the personal aspects of [his] journey with descriptions of cutting-edge research into spatial navigation and memory manipulation, as well as new ideas about how memory works." —*Scientific American Mind*

"In the tradition of Oliver Sacks's casually shrewd scientific writing, the book blends dispatches from the frontiers of science with compassionate human anecdotes. Although this exploration never shies away from formidable science and challenging psychological concepts (like contextual 'flashbulb memories,' which can be startlingly vivid and completely false), Fernyhough reinforces his lessons with elegant personal memoirs and pop-culture references. (Harry Potter, Princess Diana, and Andy Warhol all make cameos.) For a topic so elusive—discussed in methods that range from the allegorical 'crazy woman' to the brain's mysterious mechanics—Fernyhough's enthralling narrative delivers gripping insight on the way memories shape our lives." —*Apple Editor's Choice*

"His examination [is] welcoming and accessible to lay readers. His analysis is wide-ranging.... He also covers a wide swath of literary and historical ground.... A refreshingly social take on an intensely personal experience." —*Publishers Weekly*

"Outstanding. . . . Fernyhough's skills as a writer are evident both in the beautiful prose and in the way he uses literature to illustrate his argument. . . . He draws on both science and art to marvellous effect."

—*The Observer* (London)

"Combining the engaging style of a novelist with the rigor of a scientist, Charles Fernyhough has written an insightful and thought-provoking meditation on the nature of memory and its implications for our everyday lives. *Pieces of Light* will both linger in your memory and change the way you think about it."

—Daniel L. Schacter, author of *The Seven Sins of Memory*

"Tells stories to explore the deepest nature of memory, and does it beautifully. . . . In his hybrid of autobiography, journalism, and pop psychology, Fernyhough lets the stories speak for themselves to highlight memory's personal, subjective, and fragile qualities. Fernyhough takes us on a captivating journey into the mind. And he does so with great style."

—*The Telegraph* (London)

"Enlightening. . . . A crisp and knowledgeable guide to all the data that generally stays buried deep in specialist journals. Most lay people still think of memory in terms of a vast personal DVD library. . . . In fact, as Fernyhough persuasively shows, memory is far more mutable than that."

—*The Guardian* (London)

"A beautifully written, absorbing read—a fascinating journey through the latest science of memory."

—Elizabeth Loftus, distinguished professor, University of California–Irvine

"Fernyhough deftly guides us through memory's many facets. . . . Often using himself as a test case, he adds context with research and snippets from a raft of great writers. A thoughtful study of how we make sense of ourselves."

—*Nature*

"Both playful and profound, a wonderfully memorable read."

—Douwe Draaisma, author of *Why Life Speeds Up As You Get Older*

"Fernyhough weaves literature and science to expose our rich, beautiful relationship with our past and future selves."

—David Eagleman, author of *Incognito: The Secret Lives of the Brain*

PIECES OF LIGHT

ALSO BY CHARLES FERNYHOUGH

Fiction
The Auctioneer
A Box of Birds

Nonfiction
A Thousand Days of Wonder

PIECES OF LIGHT

How the New Science of Memory Illuminates
the Stories We Tell About Our Pasts

CHARLES FERNYHOUGH

HARPER PERENNIAL

NEW YORK • LONDON • TORONTO • SYDNEY • NEW DELHI • AUCKLAND

HARPER PERENNIAL

First published in Great Britain in 2012 by Profile Books Ltd.
First U.S. hardcover published in 2013 by HarperCollins Publishers.

HarperCollins books may be purchased for educational, business, or sales promotional use. For information please e-mail the Special Markets Department at SPsales@harpercollins.com.

Illustration on page 249 by Martin Lubikowski, ML Design

FIRST HARPER PERENNIAL EDITION PUBLISHED 2014.

Library of Congress Cataloging-in-Publication Data has been applied for.

ISBN 978-0-06-223790-3 (pbk.)

HB 05.12.2023

For Martha George

CONTENTS

CONTENTS

PIECES OF LIGHT

1

CASTING A LINE

"CAN YOU REMEMBER?"

It starts with a question from my seven-year-old son. We are on the grounds of our rented cottage in the Baixa Alentejo, killing time before we head to the Algarve coast for a boat trip. With his holiday money, Isaac has bought himself a handheld toy that fires little foam rockets prodigious distances up into the air, and he has lost one of them on the graveled ground behind the swimming pool. As we search, he has been chattering away about how he wants to go fishing with me when we get home from Portugal. I have told him that I used to go fishing, as a child of about his age, with my uncle in the lake in the grounds of my grandparents' house in Essex. Then, out of the blue, he asks the question:

"Can you remember the first fish you ever caught?"

I stand straight and look out at the farmland that slopes away from our hillside vantage point. I have not been fishing in thirty-five years, but my thoughts have occasionally returned to my outings with my uncle. When they do, certain images rise out of the past. I can picture the greenish lake with its little island in the middle, how mysterious and unreach-able that weeping-willowed outcrop looked to my small-scale imaginings. I can sense my jocular young uncle next to me, his stretches of silence punctuated with kindly teasing. I remember the feel of the crustless bits of white bread soaked in pond water that we used to squidge onto the fishhooks as bait, and the excitement (for a keen young amateur natural-ist) of an afternoon visitation from a stoat, scurrying along by the bull-rushes with its black-tipped tail bobbing. I remember the weird, faintly gruesome exercise of extracting the hook from a rudd's mouth and then throwing the muscular sliver back into the lake to restart its perforated life. But I have never thought about the moment of feeling the tug on the line, the thrill that prefigured the landing of a fish. And I have certainly not had the question framed like this, narrowing my remembering down to the first time it ever happened.

"I don't know," I reply. "I *think* so."

What accounts for my uncertainty? The image of lifting a fish from the water was not there in my collection of ready-made fishing memories. Because I have never (as far as I can remember) been asked this ques-tion, I have never had to come up with a corresponding memory. But I try. I ask myself: What would that first-catch moment look like? Into the well-remembered scene of the lake I insert the detail of an extended fish-ing rod, seen from the perspective of my childhood self, with something silvery dangling from the end of the line. I feel a pang of recognition, and then a shiver of boyish excitement. And then I ask myself: Did it happen? I *feel* that it did. It seems to me that the event really took place; it feels as though it belongs in the past; it comes with appropriate correspond-ing emotions; and it feels as though it happened to *me* rather than to anyone else. When I think of the memory now, a month or two after our conversation in Portugal, it has taken on an independent existence. I no longer agonize about whether it was a product of imagination, generated on demand to satisfy a small child's curiosity.

What is it like to have a memory? What *is* a memory? How is it possible to have "new" memories, like this one of catching my first fish? Have I always "had" the memory, but only just unearthed it, or have I somehow created it out of something else? What about all the other potential memories I could conjure up from that period of my life—those that are not in my consciousness right now, but that could become so, with the appropriate cues? Do I "have" them or not? What status do they have, before and after they come into my mind?

The list of questions goes on. Why did I remember this particular event and not some other? Presumably it was because my memory was clearly cued. I remembered the first fish I caught because I was specifically asked about it. But what about when a memory simply pops into my head for no apparent reason? Yesterday, for example, I suddenly had an exasperatingly random memory of the little blue-and-white-striped plastic carrier bags that were common when I was a child. We are often struck by the randomness of what we remember, and dismayed by our forgetfulness for the really important stuff. In the words of the American writer Austin O'Malley, memory is "a crazy woman that hoards colored rags and throws away food." This randomness determines what information we choose to encode about an experience, how we recall what we do actually store, and the triggers that can elicit such memories. Memories that are resistant to the ordinary processes of cueing might nevertheless be flushed out by trivial, apparently disconnected cues. Even spontaneous memories, which flash into our heads for no apparent reason, may be triggered by some subtle internal or external connections.

For all these reasons, it's impossible to answer the question of whether I "had" my first-catch memory before Isaac asked me about it. In this book, I want to show that the question is impossible because it relies on a mistaken view of what memories are.

HERE'S A MEMORY, FROM AN accomplished writer with a special interest in the topic:

It is seen from the point of view of a small person just seeing over the wall of a playground in East Hardwick Elementary School. The stone is hot, and is that kind that flakes into gold slivers. The sun is very bright. There is a tree overhead, and the leaves catch the light and are golden, and in the shade they are blue-green. Over the wall, and across the road, is a field full of daisies and butter-cups and speedwell and shepherds'-purse. On the horizon are trees with thick trunks and solid branches. The sky is very blue and the sun is huge. The child thinks: I am always going to remember this. Then she thinks: why this and not another thing? Then she thinks: what is remembering? This is the point where my self then and my self now confuse themselves into one. I know I have added to this Memory every time I have thought about it, or brought it out to look at it . . . It has got both further away and brighter, more and less "real."

The writer is the novelist A. S. Byatt. "The Memory," as she dubs it, is an example of an *autobiographical memory*, which psychologists define as those acts of remembering that relate to events and details from our own lives. You could call on anyone to recount a memory from their child-hood, and they would come up with something like this. At one level, Byatt's account illustrates the predominant view of what a memory is: a more or less stable depiction of a past event. Memories are not always as accessible as we might like—they don't always come when they are called—but they are essentially enduring representations that you carry with you, claim as your own and guard jealously. Some remember their first day at school, first kiss or first wedding day, and some don't. No one would doubt, though, that the question of whether you "possess" a par-ticular memory makes sense.

It could surely not be otherwise. Without our memories, we would be lost to ourselves, amnesiacs flailing around in a constant, unrelenting present. It is hard to imagine being able to hang on to your personal identity without a store of autobiographical memories. To attain the kind of consciousness we all enjoy, we probably rely on a capacity to make links between our past, present and future selves. Memory shapes everything that our minds do. Our perceptions are funneled by information that we laid down in the past. Our thinking relies on short-term and long-term storage of information. As many artists have noted, memory underpins imagination. Creating new artistic and intellectual works depends critically on the reshaping of what has gone before. We need our memories, and we find ways of hanging on to them. According to the conventional "possession" view of memory, we do that by filing them away in a kind of internal library, ready to be retrieved as soon as they are needed.

This view is everywhere in popular culture. In *Harry Potter and the Chamber of Secrets*, the second book of J. K. Rowling's world-famous series, Harry is threatened with having his memories "stolen," as if they were items of mental property. (If that happens, we know that Harry will stop being the person he is.) In the sixth, *Harry Potter and the Half-Blood Prince*, Voldemort's memories are capable of being accessed, distilled and transferred by Professor Dumbledore. In the hit 2009 movie *Avatar*, the hero Sully and his Na'vi comrades are able to look into Grace's memories before she dies, as though they are diary entries at which one can sneak a peek. The Internet is frequently abuzz with news stories about how scientists are coming close to targeting individual memories, confirming the impression that individual moments of experience are distributed around the brain like books in a library. Metaphors of memory are overwhelmingly physical: we talk of filing cabinets, labyrinths and photographic plates, and we use verbs such as *impress, burn* and *imprint* to describe the processes by which memories are formed.

This view of memories as physical things is guaranteed to mislead. The truth is that autobiographical memories are not possessions that you either have or do not have. They are mental constructions, created in the present moment, according to the demands of the present. Scientists try to understand this process at the cognitive level (that is, at the level of thoughts, emotions, beliefs and perceptions) and at the neural level (in terms of activations in the brain). Cognitively and neurologically speaking, Byatt does not "bring her memory out to look at it"; she constructs it anew each time she is required to do so. That is quite a different concept from the idea that a memory is a static, indivisible entity, an heirloom from the past. Rather, the view that I want to explore in this book is that a memory is more like a *habit*, a process of constructing something from its parts, in similar but subtly changing ways each time, whenever the occasion arises.

This reconstructive nature of memory can make it unreliable. The information from which an autobiographical memory is constructed may be more or less accurately stored, but it needs to be integrated according to the demands of the present moment, and errors and distortions can creep in at every stage. The end result may be vivid and convincing, but vividness does not guarantee accuracy. A coherent story about the past can sometimes only be won at the expense of the memory's correspondence to reality. Our memories of childhood, in particular, can be highly unreliable. Thinking differently about memory requires us to think differently about some of the "truths" that are closest to the core of our selves.

Novelists give us a sophisticated view of what psychologist Daniel Schacter has called the "fragile power" of memory. In her description of "The Memory," Byatt is careful to acknowledge its unreliability, malleability and deceitfulness, and the fact that it is vulnerable to a constant process of telling and retelling. She describes her awareness, even as a child, of the effort needed to construct a memory in such a way that it will not be allowed to fade: "The child thinks: I am always going to remember this." Fiction writers have much to tell us about memory, and I will be relying on their insights as I go. When they steer too close to a "possession" view of memory, however, I will look to the science of memory to set them straight.

This new, reconstructive account of memory is my real focus in this book. It is one that is largely accepted by memory scientists (with, of course, plenty of rumbling disagreements) but not yet, I think, the general population. I want to argue against the view of memories as mental DVDs stored away in some library of the mind. In fact, I would like to suggest that this mistaken "possessions" view is itself a product of the compelling storytelling (and restless search for psychological causes and effects) with which our brains are constantly busy. I want to persuade you that when you have a memory, you don't retrieve something that already exists, fully formed—you create something new. Memory is about the present as much as it is about the past. A memory is made in the moment, and collapses back into its constituent elements as soon as it is no longer required. Remembering happens in the present tense. It requires the precise coordination of a suite of cognitive processes, shared among many other mental functions and distributed across different regions of the brain. This is how Schacter, one of the pioneers of the approach, sums it up:

> We now know that we do not record our experiences the way a camera records them. Our memories work differently. We extract key elements from our experiences and store them. We then re-create or reconstruct our experiences rather than retrieve copies of them. Sometimes, in the process of reconstructing we add on feelings, beliefs, or even knowledge we obtained after the experience. In other words, we bias our memories of the past by attributing to them emotions or knowledge we acquired after the event.

This is a very different view of memory from the one that I think most non-psychologists hold. Understanding how it emerged involves taking a fascinating journey into the science of how we are shaped by our pasts.

FOR A LONG TIME, AUTOBIOGRAPHICAL memory was not a topic that appealed very much to me. As a psychology undergraduate in the late

1980s, I was interested in those details of mind and behavior that would submit themselves to formal analysis. Memory was too unmeasurable, too unreliable, too subjective, too fuzzed up with messy human detail. Everyone remembers the past differently, because everyone lives it differently. It was hard to know how to make a science out of memories, and I was drawn instead to questions where the answers were more quantifiable. I wanted to get scientific about hard numbers (which I thought, at the time, was the only way of getting scientific), and all memory seemed to offer me was personal stories.

Now, as someone who divides his time between scientific psychology and the writing of fiction and nonfiction, these are precisely the qualities of autobiographical memory that appeal to me most. I am interested in it for some of the same reasons that a novelist might be: because it gives the richest illustration of the complex ways in which human beings make sense of their own existence. The painstaking work of generations of memory scientists has illustrated those interactions between different cognitive systems that underpin even the most ordinary act of remembering. To have any chance of being later recalled as an autobiographical memory, the details of an episode must be encoded, stored, labeled and eventually retrieved. They must make connections with areas of the brain subserving sensory perception, navigation, emotion and consciousness. Above all, they must be stitched together by a sometimes effortful process of imaginative reconstruction.

None of this would be possible unless the rememberer had a sense of his own self as unfolding through time. In my last book, I traced the emergence of this self-understanding in the case of my own young daughter, Athena. A theme that emerged for (and rather surprised) me in writing that book was the impressive effort of the young child to make sense of her experience in terms of a narrative. In this book, I am going to pick up and build on this theme. I want to explore how an ability to move mentally through time underpins both the looking backward of autobiographical memory and the projections into the unknown involved in future-oriented thinking. To do this, I am going to focus on human stories. By letting memories speak through narratives, I hope to expose some common myths about how memory works.

I am not the only one who is getting interested in memory again. It is arguably a basic human need to try to understand one's past and create a coherent narrative about where one has come from. Findings that many of our cherished memories may well be inventions, therefore, seem to challenge our sense of identity in potentially catastrophic ways. Some of the most powerful and influential artworks of recent times have been concerned with the deceptions of autobiographical memory: W. G. Sebald's genre-defying novel *Austerlitz*, for example, or Christopher Nolan's 2000 film *Memento*. Memoir is an increasingly popular literary genre, and yet it rarely examines its own workings in the sense of asking whether the memoirist should trust his or her recollections.

Many of us feel that our memories let us down, and scores of self-help books promise to help us improve our remembering. Loss of memory can be a sign of encroaching dementia, and our interest in improving our memory must tap into our anxieties about Alzheimer's disease. On the other hand, some people remember too much. For those affected by trauma, remembering can be a vicious cycle leading to crippling psychiatric problems. And the foibles of memory can have desperately important implications when it comes to witnesses and victims remembering events in court. The work of the American psychologist Elizabeth Loftus and others has shown that memories are very susceptible to being distorted by information provided after the event, and that in certain conditions it is even possible to "implant" memories just by giving people appropriately suggestive information. The evidence that people can vividly remember events that never happened must make us rethink our emphasis on eyewitness testimony in legal proceedings.

Too often, though, the fallibilities of memory are insufficiently acknowledged. Even supposed psychology experts can turn out to be not much better informed than the general public about how memory works, as one recent study of Norwegian psychologists showed. Around 850 psychologists were presented with twelve statements about memory, and asked whether they agreed or disagreed with them. For example, one statement read: "At trial, an eyewitness's confidence is a good predictor of his or her accuracy in identifying the defendant as the perpetrator of the crime." Respondents' answers were then compared with those that were considered to be the "correct" answers according to current scientific knowledge. The psychologists got an average of 63 percent "correct" (compared with 56 percent for members of the general public). A link to the test items and the correct responses is given in the notes at the back of this book.

If you didn't do so well on these questions, you are in good company. Another recent study involving a telephone survey of a large sample of ordinary Americans asked people whether they agreed with six statements chosen to conflict with the expert consensus. Topics included amnesia and identity, confidence in testimony, the analogy between memory and video cameras, the influence of hypnosis on memory, attention to unexpected objects and the permanence of memory. Considerable proportions (in two cases, substantial majorities) of people agreed with the false statements. For example, 83 percent of people thought that amnesia resulted in an inability to remember one's own identity, and 63 percent thought that memory works like a video camera.

It seems that we get memory very wrong. And yet, when the topic appears in the media, the public appetite for information seems voracious. The American journalist Joshua Foer reportedly gained a seven-figure advance from a publisher for his study of the "mental athletes" who compete in memory contests. As I write this, in January 2012, an edition of *Scientific American Mind* is overturning some common myths about memory and forgetting, while a special supplement of the *Guardian* shows how we can make the most of our memory's power. An online memory experiment accompanying the *Guardian*'s supplement was visited by more than 80,000 people worldwide, while an earlier Web survey for the BBC generated much controversy, particularly concerning the authenticity of preverbal childhood memories.

This interest in memory is part of a spreading fascination with the often counterintuitive discoveries of modern psychology and neuroscience. We are now used to reading about research that calls into question deeply held assumptions about how our minds work. We know that there is not one single center of experience in the human brain; we are told by scientists that our minds are ragtag collections of semi-independent processors, each evolved to do a specialized task. We know that when we look out at a visual scene, we don't actually see the scene in its entirety; we see fragments that are later stitched together to create the illusion of a unified scene. Memory doesn't stand out in this respect, if you consider it alongside the other fragmentary kinds of cognition with which our brains are constantly busy.

That said, the study of memory does pose some very specific challenges. My undergraduate pessimism about whether it was possible to have a science of personal stories is still grounded in some real uncertainties. Asking people about their memories is fraught with difficulties. Memories are changed by the very process of reconstructing them, and every memory that an experimental participant reports is likely to have been contaminated by previous acts of remembering.

But scientists have found ways to study autobiographical memory, and have been doing so systematically for over a hundred years. Beginning with the pioneering (and very different) self-examinations of memory conducted in the 1870s and 1880s by Sir Francis Galton in England and Hermann Ebbinghaus in Germany, memory researchers have subjected their participants to tests of recall for nonsense syllables, interviews about their earliest memories, and experiments on the power of sensory stimuli, such as music and smells, to trigger recollection. The reconstructive view of memory has its origins in the work of Sir Frederic Bartlett, the first professor of experimental psychology at the University of Cambridge, whose most famous work was summarized in his

1932 book *Remembering*. Bartlett asked his participants to read a North American Indian folktale called "The War of the Ghosts," involving a battle between ghostly warriors, and then to retell the story under a range of different conditions. He found that people's memory of the story was affected by their own beliefs about how the world worked, and that they distorted the story to fit their own knowledge structures, leaving out bits that seemed to them irrelevant and changing the emphasis and structure of the story to fit their own understanding. Bartlett concluded that our memory of events reflects the information we encoded at the time, mixed up with inferences based on all sorts of other bits of knowledge, expectation and belief.

The modern inheritors of Bartlett's reconstructive view of remembering are researchers like Daniel Schacter, Elizabeth Loftus, Endel Tulving, Donna Rose Addis, Antonio Damasio and Martin Conway. Drawing on a distinction made by the philosopher Bertrand Russell, Conway has distinguished between two forces in human memory: the force of correspondence, which captures memory's need to stay true to the facts of what happened, and the force of coherence, which works to make memory consistent with our current goals and our images and beliefs about our own selves. Memory is an artist as much as it is a scientist. Among those who study it scientifically, the conventional view of autobiographical memory has been upended by one in which memories are constructed, through a process that combines stored sensory and emotional information with more formal and schematic descriptions of knowledge about one's past life, and that requires the simultaneous functioning of many different cognitive systems.

Memory means different things to psychologists. Autobiographical memory is an interesting case because it straddles the most basic of the distinctions that scientists make between types of memory: that between *semantic memory* (memory for facts) and *episodic memory* (memory for events). Our memory for the events of our own lives involves the integration of details of what happened (episodic memory) with long-term

knowledge about the facts of our lives (a kind of autobiographical semantic memory). Another important distinction is that between *explicit* or
declarative memory (in which the contents of memory are accessible to
consciousness) and *implicit* or *non-declarative* memory (which is unconscious). As we will see, this distinction is particularly important when it
comes to the question of how memory is affected by trauma and extreme
emotion.

Autobiographical memory is also a form of long-term memory, and
so I will not be saying much about the world of short-term memory, or
working memory as it is more commonly known. None of these varieties
of memory is unitary and freestanding, but rather all depend on several
different cognitive systems and neural pathways. Implicit memory, for
example, relies on different neural circuits than autobiographical memory. When you learn a new motor skill, your cerebellum (tucked away
inside your skull behind the nape of your neck) buzzes into action. When
you take a wrong path out of habit, you are seeing the evidence of information patterns stored in your basal ganglia, situated deep down in the
middle of the brain above the brain stem.

When it comes to autobiographical memory, it is a mistake to think
that memory traces are stored in any one part of the brain. Indeed,
the search by early memory researchers for what became known
as the *engram*—the single bioelectrical trace that a memory leaves in the
brain—was always destined to end in failure. Although not my focus in
this book, much progress has been made in understanding the process
of *long-term potentiation*, the structural changes in neurons that underlie the brain's storage of information. You will doubtless have seen news
stories about how "memories" are formed when certain chemical changes
happen in the synapses of your brain's nerve cells. Although this research
is fascinating and important, these are not memories as I am interested
in them: they are about individual cells, not human beings. They work at
a different level of explanation.

A similar point can be made about the process of *reconsolidation*, according to which "memories" are reformed at the molecular level each time they are activated. Reconsolidation became a hot topic in memory science after researchers at New York University used a chemical proven to disrupt the formation of memory traces in rats that had learned to avoid an electric shock. The big discovery was that the chemical (a protein inhibitor) was also effective at the *recall* of a memory (when rats were remembering the electric shock) as well as at its initial encoding. If the memory trace had been permanent, this should not have happened. Instead, the data showed that a memory trace can be altered after the event, in the absence of the original stimulus. Reconsolidation seems to point to a molecular mechanism through which memories can be changed by subsequent events. But it does not show *how* they are changed. For that we also need to investigate memory at the cognitive level, that is, at the level of individual people's thoughts, beliefs and biases.

A similar note of caution needs to be sounded about other neuroscientific findings. The new science of neuroimaging offers an entirely new perspective on the age-old question of where in the brain our memories, and therefore in some sense our selves, reside. Memory scientists have studied the remembering brain through neuroimaging scans, electroencephalography (EEG) experiments and the careful interviewing of brain-damaged patients. Brain imaging shows activity in the frontal lobes, where the efforts to reconstitute a remembered experience are initiated, through the emotional circuits of the amygdala system and the associative centers of the neocortex, to the occipital lobe at the back of the brain, where the characteristically visual qualities of autobiographical memories are stored as sensory fragments.

Understanding these neuroanatomical patterns is very valuable. If we are to have a science of human experience, we have to tackle it on different levels, which would include at least the molecular, neural, cognitive and social. And the study of brain processes of learning and memory has contributed a great deal to our understanding of how memory works. For that reason, I will be referring in particular to new research in cognitive neuroscience, the discipline that integrates findings from experimental psychology, neuroimaging and neuropsychology (studies of brain damage). The neural tentacles of memory spread far and wide, and many different brain systems are involved. To set the scene, here's a brief overview of the main brain areas that will come into focus (see diagram, p. 249).

If you place a finger above your ear and imagine being able to push it in about two inches, your virtual fingertip would touch on the single most important structure for autobiographical memory. Over the decades that it has been studied, the hippocampus has been implicated in psychological processes as diverse as memory, spatial navigation and anxiety. Often likened to a seahorse because of its curved, flowing shape, it sits at the center of a network of memory circuits in the medial temporal lobes (you have a hippocampus, and a medial temporal lobe, on each side of the brain). The hippocampus works closely with areas of cortex nearby, known as the perirhinal and parahippocampal cortices, which sit just under the hippocampus at the front and back respectively. This relatively small area of the brain extends into a wider network of memory-related regions, including the amygdala, which is connected to the front of the hippocampus and is crucial for learning about the emotional significance of stimuli. Moving farther toward the front of the brain, the medial temporal lobe memory circuits connect with the control systems of the prefrontal cortex. At the back of the brain, the occipital cortex stores the visual perceptual details that are so important in autobiographical memory.

There is, of course, more to remembering than neural systems. I think that if we are really to unpick the mysteries of memory, we need to put the story back into the science. One of my aims in this book is to capture the first-person nature of memory, the rememberer's capacity to reinhabit the recalled moment and experience it again from the inside. The great memory scientist Endel Tulving called this quality of memory *autonoetic consciousness,* and explaining it is one of the biggest challenges for memory researchers. The scientific need for replicable experimental findings has meant that the personal, subjective quality of memory has often been ignored, although this tendency has begun to be redressed in recent years, with a new movement toward exploring the qualitative and the narrative. Memory researchers now spend more time getting to know their participants' individual stories, whether they concern the beguiling confabulations spun by those whose memory systems have failed them, or the sensually rich "first memories" produced when people are interviewed about their very early childhoods. I want to do the same thing, letting the stories speak for themselves in illustrating the fragile and complex truths of memory.

I start my journey by getting lost. Returning to a city that I used to know very well, and trying to find my way through once-familiar streets, I am given a persuasive lesson in how memories are mediated by previous acts of remembering. Finding your way in a landscape requires that you have accurate memories of where you have been, but it also depends on your ability to encode knowledge about space and time. I explore how this kind of information is processed by the hippocampus, ultimately producing an internal map of one's location in the terrain. When you are lost, as I am when I return to another once-familiar city, these maps go awry. I ask what the workings of this kind of geographical amnesia tell us about how memory operates, in landscapes of the imagination as well as of reality. My wanderings in my old home cities demonstrate that we can get lost in our pasts in the same way that we get lost in an unfamiliar terrain.

I then look at the role of the senses in autobiographical memory. Writers as diverse as Marcel Proust and Andy Warhol have been eloquent in describing the power of sensory stimulation to unlock the past. Smells and music are known to be strong triggers for involuntary memories, and I ask whether there is anything special about these sensory modalities that make them particularly effective at unlocking memories. These examples show how making autobiographical memories is intimately linked to our sensory and emotional experience of the world, and demonstrate how memory depends on seamless collaborations between many different cognitive and neural systems.

Memory's complex synergy of cognitive and neurological functions must take time to develop. Infants and small children remember things, but they need to achieve certain milestones before they are able to do genuine autobiographical memory: to put themselves at the center of the events they are describing. Asking when memory gets started tells us a great deal about the different psychological capacities that allow us to trace our own selves back into the past. In Chapter 4, I look at why we mostly cannot recall our childhoods, and why our earliest memories are so full of rich sensory detail.

In Chapter 5, I return to a landscape from my own childhood, for a lesson in how your memories of a person can be unlocked by a return to the places you used to share. It is a well-established finding that we are better at remembering events and information when we are asked to recall them in the same context in which we laid the memories down. I look at how memory attunes itself to the meaning of information rather than its surface details, and examine how memories for events are framed by their context, so that having a memory is a process of matching the cues that are present at retrieval with the information that was encoded at the time.

One striking fact about childhood memories is that they are built up through collaborative acts of recollection with parents and other care-givers. As I describe in Chapter 6, talking together about the past seems to be vitally important in children's creation of a self that extends through time. In adulthood, memory can be something that has to be negotiated socially. The idea that the past is a story that we tell ourselves, whose viv-idness can be no guarantee of its authenticity, highlights our reliance on language for social acts of remembering. If our autobiographical memory system serves to create a coherent narrative of our own past, it is a system that can frequently fool us into believing stories that are not true, as evi-denced by the fact that many of us "remember" events that we no longer believe actually happened.

In Chapter 7, I ask what memory is for. Doing good psychology has always involved taking an evolutionary perspective, and the study of memory is no exception. In fact, there are some very good reasons for thinking that memory evolved not to keep a record of what has happened but to predict what will come next. If memory is fallible and prone to reconstructive errors, that may be because it is oriented toward the future at least as much as toward the past. Some of the most exciting recent research in this area has shown that similar neural systems are involved in both autobiographical memory and future thinking, and that they both rely on a form of imagination.

If our memories are constructions of an imaginative process, we nevertheless need some way of keeping track of what mental experiences actually happened to us in the past, as opposed to events that we have simply imagined. In Chapter 8, I explore the feelings that tell us when we are remembering. It turns out that one of the biggest challenges of memory is keeping track of the source of our experiences, and some of the most distinctive memory errors occur when we fail to distinguish what we have remembered from what we have merely imagined.

More than anything, memory is a great storyteller. Not only do we stretch our narrative capacities to the limit when we construct an autobiographical memory, but we also eagerly spin tales whenever our memories leave us with gaps in the record. In Chapter 9, I look at what we have learned about autobiographical memory from the study of patients with brain damage. I meet a woman who has lost the ability to form new memories, and hear the story of another who lives his life in a state of continual déjà vu. In both cases, incongruous experiences of remembering can lead the sufferer to create elaborate stories, or confabulations, around them.

The subject of Chapter 10 is memory for trauma. I meet a man whose life was ripped apart by a tragic accident, and ask whether traumatic memories function in the same way as non-traumatic ones. When I talk to Colin about his treatment for posttraumatic stress disorder, I realize that the crucial thing seems to be the way in which fragments of remembered experience are integrated into a coherent whole by the parts of the brain that are involved in stitching together autobiographical memories.

The goal of therapy in such cases is not forgetting, but a more accurate, inclusive and unbiased remembering.

In Chapter 11, I look at memory in old age. None of us is immune to the *reminiscence effect,* the phenomenon whereby events from one's late teens and early twenties stick in memory better than anything else. In the case of my ninety-three-year-old grandmother, the memorable events from her life happened in the 1930s, when she was a teenager growing up in the Jewish East End of London. I ponder the effect of this continual salience of one's youth on a mind that has outlived those years by many decades. I ask how remembering is dependent on finding a match between the language used when the events were encoded and that used at retrieval. I ask why life speeds up as we get older, and why (paradoxically) time also drags. The theme of this chapter is reminiscence: the act of recollection on demand, the effortful process of casting back to the past. But it is also about some of the particular qualities that distinguish remembering in old age.

I end by thinking about the future of memory. Some of the remaining mysteries of memory look set to be solved by recent breakthroughs in molecular science and neuroimaging. Others will probably continue to puzzle us for decades to come. The more we learn about our memories, the greater our opportunities for manipulating them, changing them and possibly even getting rid of them. The ethical implications of understanding memory may be more far-reaching than we can currently imagine. I ask how collections of people, even entire cultures, can "remember," and what happens when memory becomes politicized. In another context, that of law and eyewitness testimony, the scientifically acknowledged frailty of human memory is beginning to be factored into legal judgments. I end by considering memory as a form of knowledge with a disputed status, serving the self as much as it serves truth.

2

GETTING LOST

THE GREAT CITIES break the rules of memory. A place you have never visited can seem rich with remembered experience, if only because so many others have been there and encoded its sights before you. Before you set foot on the streets of a city like Cambridge, you are hooked into a fictional narrative. All those bescarved undergraduates, their bicycles tinkling over cobblestones: they are scenes from other people's memories, rehashed and polished to artifice through movies, novels and impossibly perfect college brochures. The city has been remembered even before you get there.

That was certainly my feeling when I started there as an undergraduate in the mid-1980s. The place felt like a work of fiction, a movie outtake about a lifestyle that should long have been declared extinct. I was pre-jaded, already bored with what I had never actually experienced. I resented the city's unreality, the fact that people were doing what had been brainlessly preordained for them. I sat in my modern college room and fumed about the posh, noisy families punting on the river below my window. It seemed inconceivable that they could not recognize what clichés they had become. Looking back, I realize that this was more my problem than theirs. I have always had a childish squeamishness about being seen to be doing the obvious thing. I would rather be spotted in a bookshop queue holding a guide to medieval beekeeping than preparing to splash my cash on a copy of the latest bestseller. Arriving in Cambridge and heading straight to the punt stand seemed, to my precious adolescent sensibilities, the worst kind of unexamined living.

Over time, of course, the fictional Cambridge gave way to a real one. The city ended up being my home for seven years: three of them spent living there as an undergraduate and then, after a gap year's traveling, four years of doctoral research. The heritage movie set that confronted me in 1986 became a place I knew as well as any other. I got drunk there, fell in love there, achieved some academic success. I stayed there for a long time. Whatever preconceptions I started out with became overwritten pretty quickly with "real" memories. It was soon a place I was genuinely familiar with.

Then I moved away. In the seventeen years since I left, I have gone back only three or four times. When I think about Cambridge now, it simmers in my past like an experiment in memory. I'm curious to know how well I recall this city of my formative years. I want to know how you become attached to a place, and how long those attachments last. But I also suspect that thinking about memory for places will tell me some things about how memory works in general.

Take one basic idea, for example. You remember stimuli that you are exposed to. Cambridge was not only the city where I fumbled my way into adulthood; it was also the place where I first studied human memory. I was fortunate enough to be taught by Alan Baddeley, the great English pioneer of memory research. In one lecture, Baddeley asked us to imagine a one-penny piece. None of us had any problem in conjuring up an image of the thing, but what came next was more of a challenge. Could we recall the design on the tails side of the coin? None of us could. Despite the fact that we must each have handled this unit of currency many thousands of times, we had been left with no memory of its features. Simply being exposed to a stimulus does not guarantee that you will remember it. You need to act on the stimulus in some way; you need to process it. Remembering depends on many factors, but at the very least it requires attention.

I surely paid attention to the streets around Downing Street and Tennis Court Road. My undergraduate degree was in Natural Sciences, which meant that I walked this way from my college most days—including Saturdays, when we had nine o'clock lectures. My abiding image is of hordes of young men (they were all men) on racing bikes, pedaling along the straight quarter mile of Tennis Court Road as fast as they could. It was risible and slightly frightening. We bipedal slackers used to shout sarcastic encouragement, shading to outright abuse, at the geeky Natural Scientists as they went hurtling past. Cruelly likening it to the famous annual bike race in France, we used to refer to it as the Tour de NatSci. I mention this because I know it is a genuine memory: I have not been in that part of Cambridge at that time in the morning since my undergraduate days, and so I can't be recalling some more recent intervening experience. Like the police riot vans that gathered on Bridge Street on Saturday nights, it is not the sort of tourist brochure image that we are otherwise bullied into remembering about Cambridge. It is in my memory because I saw it. There is no trickery here.

As soon as that image of the hurtling scientists takes shape in my memory, a whole host of other connections start being made. I trust the image; I don't always trust its baggage. If I was walking up Tennis Court Road, I must have been heading for the Chemistry Department on Lensfield Road. One of the things that made first-year chemistry so horrible was that we had Saturday-morning lectures. I remember that as a fact; I can confirm it with friends who were there. As soon as I do that, I am leaving the realms of what I actually recollect, and I am starting to bring in factual knowledge about events in my life. The image is not tagged "Saturday morning, Tennis Court Road," like a picture on a photo-sharing site; it is just an image. But I can pinpoint it to a time and place by bringing in relevant information about what I was doing there.

Some other impressions cling to the image as well. I remember being hungover: not an unusual feeling for me at that time, but a particularly intense one at ten to nine on a Saturday morning after a big night in the college bar. The medieval sandstone of the college buildings links in my memory to a particular woozy, lovelorn restlessness. There was someone I wanted who didn't want me. The Smiths had written a song about it. What amazes me is that the feeling isn't just "out there," floating in the ether; it is steeped into the fabric of Pembroke College and the buildings at the back of the old Addenbrooke's Hospital. The city remembers my pain, or else I remember it remembering it.

But none of this should surprise me. We are familiar with the idea that a return to a childhood home can bring back memories. You walk through the door of that house you lived in with Mum and Dad and—whoosh!—the memories come flooding back. But something even more powerful can apply to the places we live in as young adults. Memory researchers tell us that events from our late teens and early adulthood have a particular hold on our memories. If you ask an adult to recall episodes from their past, and then arrange the remembered events according to the age at which they occurred, you are likely to see a peak around the age of twenty. Researchers refer to this peak as the *reminiscence bump*, and they have put forward various explanations to explain its cause. One account holds that early adulthood is remembered best because that's when the most significant events happen to you. Leaving home, falling in love, breaking up with a partner: these are all emotionally intense experiences that stick in the mind, and they tend to be experienced first in early adulthood.

In memory, then, a university town should be a special place. You arrive there at an impressionable age, and everything seems strange and new. You live there for a few years, possibly as intensely as you will ever live anywhere. And then you move on. Because your friends usually move on too, and because you are not tied to the place by family or work, there is often little reason to return. Your memories of the place are sealed off, undisturbed. Whatever you recall can be recalled purely, untainted by recollections of later visits. When I think about how it figures in my own life narrative, Cambridge should be richer with memories than any other place I have lived.

Puzzling, then, that my arrival should fill me more with confusion than with anything else. Walking up St. Andrew's Street toward the city center on this warm July day, I am confronted by sights that are both known and strange at the same time. The city in which I lived for seven years as a student is undoubtedly familiar, but its sights are also conspiring to make me feel utterly lost. I realize that Cambridge went through an economic boom in the years after I left, and much has changed about its physical fabric. But even the carefully preserved college buildings look indefinably strange. I glance in one direction along the street and the scene has a shimmering, picture-postcard-ish unreality; I glance back, a moment later, and it is already familiar. My problem is that I can't stop my brain from processing this information in its usual way. I can't say, "Hold it right there; give me nothing of this except the connections with the old memories." I see a sight and I encode it. I am relearning even as I try to remember.

I find my way to the college where I am staying and dump my stuff. I am in a student room in Sidney Sussex, and my next goal is to get myself across town to the English Department on Sidgwick Avenue. Since this is the city I supposedly once knew better than any other, I decide to take a shortcut down Trinity Lane, a pedestrian route that slips between the colleges down to the river. I dimly remember that I need to turn off here to find the modern bridge that crosses the Cam, but I don't know which turning I want to take. One vista looks vaguely familiar, a narrow alleyway that runs between the backs of college buildings with a heavy wooden gate at the end. But it yells "dead end," so I carry on along the same lane. When I find myself emerging onto the main drag of King's

Parade, I realize that I must have gone wrong. I head back along the lane to my previous point of indecision, and the narrow alleyway now looks overwhelmingly familiar. I *know* this alleyway; I know it with all my heart. It is as though I have finally succeeded in retrieving a particular deeply buried memory: familiarity was lurking there, just under the surface, but it needed a bit of coaxing to make it reemerge. But then the alleyway *should* feel familiar, because I saw it only a moment ago. Perhaps I am simply recognizing the new experience of a moment before, and getting the source of the memory wrong: experiencing it as an old memory when it is actually a brand-new one. It is an unsettling feeling. How can I ever distinguish what I know from what I do not know? How can I judge what is a "real" memory, dating back to my undergraduate days (and only now pulled, with some effort, back into the light), and what has only become familiar in the short time that I have been back here?

I am in Cambridge for a multidisciplinary conference entitled "Memory Maps," at which I will be giving a talk on how young children begin to navigate the landscape of memory. One of the speakers is Rebecca Solnit, a San Francisco–based writer whose latest book is a cultural history of losing your way. In *A Field Guide to Getting Lost*, Solnit celebrates the human predilection for erring from the path. Getting lost can be an event over which we can exert some voluntary control. We can surrender ourselves to the unfamiliarity of a new environment, and delight at the new avenues (topographical and psychological) that it opens up. Or we can lose our way accidentally, unhappily, even dangerously. We can play with the boundaries of lostness, as you might do in an unvisited but famous city, strolling around with no particular direction in mind, but aware that you are likely to come across a famous landmark, or that you have a tourist map to help you out if necessary. For Solnit, *lost* has two different meanings. "Losing things," she writes, "is about the familiar falling away, getting lost is about the unfamiliar appearing."

Whether it is actively sought or not, getting lost is at some level about the success or failure of memory. Finding your way requires you to pay attention to, encode and retrieve information about where you are going and where you have recently (or not so recently) been. It asks you to construct a mental map of your surroundings, and keep a note of your own position in it. It also requires you to keep track of how well you are doing. Solnit notes that people who get lost often aren't paying attention when they do so, so they don't know what to do when they realize their predicament, and won't admit to themselves that they don't know. "There's an art," she writes, "to attending to weather, to the route you take, to the landmarks along the way, to how if you turn around you can see how different the journey back looks from the journey out . . . The lost are often illiterate in this language that is the language of the earth itself, or don't stop to read it."

Not everyone knows when they are lost. When they know it, they will not always admit it. I'm fascinated by how this affects what people do when they cannot find their way. Solnit tells the story of a terminally ill child who got lost during a game of hide-and-seek on an outdoor adventure course. When he could not be found, a search-and-rescue team was sent out in the freezing night to look for him, fearing the worst. At dawn, the rescuers heard the child's feeble whistle. Rather than wandering off into the unknown, he had curled up into the gap between two trees and waited for sunrise. He knew he was lost, and that his only hope lay with other people. The child's life was eventually cut short by his illness, but this story at least had a happy ending. Children are often particularly willing to admit that they don't know where they are, and this consciousness helps to keep them safe. They recognize that they are in trouble; they settle down and wait for help to come to them.

Solnit's story is recounted by a ranger, Sallie, who works in the dangerously sublime terrain of the US Rockies. Those who work with the lost for a living—park rangers, coast guards, mountain rescuers—know that people act in particular ways when they don't know where they are. They have instincts about where lost walkers will turn up. One hunter friend of Sallie's tells Solnit about how her husband had driven his snowmobile straight to the point where a lost doctor lay freezing, knowing by hunch where an off-trail adventure might have gone wrong. It seems that those who encounter the wild on a daily basis build up a body of expertise on what people do when they are lost, and where to head in order to find them.

Getting lost is a telling kind of amnesia. It reminds us that we rely on memory all the time for finding our way in the world, for physically navigating the spaces we move through. It shows us that we are not disembodied minds, churning out computations within a space of pure information, but that we are always engaged with a physical world. One of the ways in which we encode information about space is relative to our own bodies. The directions *left* and *right*, for example, are determined in relation to our own position, referring to different absolute directions depending on where our bodies are facing. Something that's on your left on the way out will be on your right on the way home. When there are no landmarks to navigate by, you are forced to make sense of a landscape relative to your own body, which is of course constantly moving and changing orientation. Where you really want to be able to rely on the certainties of *north* and *east*, you're stuck with *left*, *right* and *straight ahead*. Scientists of navigation refer to this kind of coding of space as *egocentric*, since it is determined relative to the self. They contrast it with *allocentric* coding, which treats space as independent of one's own position.

Reliance on egocentric coding might account for some of the distinctive concomitants of getting lost. It is a familiar idea that someone lost in a desert or other featureless region will walk in circles. In Mark Twain's semiautobiographical collection of traveling tales, *Roughing It*, for example, the narrator and his companions unwittingly follow their own tracks through the snow for more than two hours before realizing their mistake. Wandering the Forest on the trail of a suspected Woozle, Pooh and Piglet walk in circles until they realize that the Woozle's tracks are in fact their own. A recent study found empirical support for the idea that people do indeed walk in circles when they are lost. Volunteers were equipped with a portable GPS system and set walking in either a German forest or a part of the Tunisian Sahara, with instructions to try to follow a certain direction. When the participants were able to see the sun, they were pretty good at keeping going in a straight line. When they had no such external reference point, they walked in circles or made odd, chaotic digressions. One participant was asked to walk in the desert at night, and only veered off course when the moon became obscured by clouds. Those who had no heavenly body to steer by behaved quite similarly to another group of participants who were observed walking blindfolded in an open field. Rejecting explanations to do with differing leg sizes or asymmetries in the brain, the researchers concluded that we can rely on our balance and the sense of our own body to keep us straight over short distances. Over longer distances, however, random errors build up in our sensory system to send us off course.

Most of the time we are not lost in featureless landscapes or bumbling along in the dark. People face navigational challenges in environments where there are plenty of landmarks to steer by. Finding your way in such a landscape requires that you have accurate memories of where you have been and when, but it also depends on your skills in allocentric and egocentric coding. Like much in the world of memory, asking how we remember spatial information brings us straight back to the hippocampus. It has been known since the early 1970s that special cells in the hippocampus of the rat fire when the animal is in a particular location. When you put the animal back in an environment it has been in before, the same *place cells* fire, suggesting that specific patterns of place cell activation "remember" the spatial context the animal has experienced. With more recent discoveries, these place cells have been joined in the roster of hippocampal navigation equipment by *head-direction cells* in the parahippocampal cortex, which fire when the animal is pointing in a specific direction and act as a kind of internal compass. Together, these cells form the neural basis of remembering allocentric and egocentric information about space.

A third class of cell was discovered in 2005 by Norwegian researchers, causing great excitement among neuroscientists of navigation. By
recording the outputs of cells in the dorsomedial entorhinal cortex, a part
of the brain that closely abuts the hippocampus and forms one of its most
important inputs, the researchers showed that certain cells, which they
dubbed *grid cells,* fired whenever the animal was at the same point on
any of a grid of hexagons representing the space in question. The theory
is that grid cells divide up a navigated space into equal-sized units, like
the squares on an Ordnance Survey map. To see how this might work,
imagine carving up the space that surrounds you now into just such a
hexagonal honeycomb. A grid cell might fire every time you stood at the
top corner of one of these hexagons—and it would fire no matter where
on the grid you were standing, as long as you were in the same part of
the hexagon. Grid cells therefore seem to code for information that is not
about absolute space, nor simply about the direction in which you are
pointing. When different grid systems are overlaid on top of each other,
they can provide very detailed information about an animal's location.
The hexagonal grids represented by grid cells can also provide information about how far the animal has traveled, just as sailors can use lines of
latitude and longitude in finding their way around the globe.

The exciting thing about grid cells is that they suggest how disparate
bits of spatial information—place, distance and direction—are integrated
into a mental map of the world. The kind of neuroscience that allowed the
discovery of grid cells, namely electrical recording from single neurons,
is not usually possible in humans, and so it was an exciting development
when a brain imaging study using functional magnetic resonance imaging (fMRI) suggested that similar hexagonal grids are represented in the
human entorhinal cortex. Participants were asked to navigate a virtual
reality arena, in a task that was designed to be as close as possible to
that performed by rats in a connected study. The researchers were able to
show that the same grid cell pattern of activation occurred in the human
brains as would have been predicted from the rat data. Although specific

cell locations have not yet been pinpointed, these findings suggest that grid cells exist in humans and that they operate in similar ways to those in the rat brain.

The involvement of the hippocampus and its related cortical structures does not end there. The latest fMRI research suggests that certain cells in the human hippocampus fire at rates proportional to one's distance from a goal: the distance as the crow flies, as well as the distance you would actually have to travel (along paths and around obstacles) to reach that point in space. Studies of the rhythmic firing of hippocampal cells also show that there is a particular low-frequency oscillation, known as the theta oscillation, that permeates the hippocampal system. It is thought that hippocampal theta may provide the timekeeping signal that allows us to integrate information about where we are in space with details of how long we have been moving for. As any sixteenth-century navigator would have told you, you need more than a compass to find your way around the globe; you also need some way of keeping track of time. It's possible that the theta rhythm provides the time signal that, combined with the computations of head-direction, grid and place cells, can give us the information we need for keeping track of our wanderings. Although there is still much to be done in working out exactly how this might happen, it has even been proposed that these specialized systems may provide the basic timing information for the "what happened and when" quality of autobiographical memory.

When everything is working smoothly, the navigation systems of the hippocampal region construct a system of representations of where you are, a mental map of your past and present locations as well as your ultimate goal. Getting lost means being unable to connect that internal map to what you perceive of the world around you. It means going wrong, and then going wrong again because you have already gone wrong before. When you are lost, you are messing up your delicate neural map even as you are constructing it. And the vagaries of autobiographical memory, those time slices of experience that seem only tangentially connected with the representation of space, combine to make these problems even more acute.

SYDNEY IS ANOTHER CITY THAT I know very well, but that can also trick me with misleading versions of itself. When I lived here, I used to say that I knew my way around this beautiful sprawl of a city better than anywhere on earth. But a little knowledge can be a dangerous thing, and that is never so true as in the mazes of remembering.

It is a bright winter's day, and I am crossing Sydney Harbor with my mind on another experiment in memory. My plan is to get off the ferry at Cremorne Point, one of the fingers of land that jut down into the harbor from the North Shore, and walk down the jetty to the concrete concourse where the buses turn. The little shop where I used to buy ferry tickets is unpainted and padlocked. I climb the steps to the Reserve, a lush green outcrop of bushland with breathtaking views back to the city. The track has been resealed, and new information plaques spell out the details of the view and the history of this miniature country park. My eyes catch on an empty kids' playground. I would not have noticed it last time. You process what is salient to you, and opportunities for entertaining toddlers were not relevant to me when I last walked along here. It is late morning, and the only people around are a promenading retiree couple and a solitary middle-aged jogger wearing sunglasses against the winter sun. In a house being renovated behind me, a builder shouts above the whine of a power tool. I can hear the sound of a strimmer from across a yacht-strewn Mosman Bay.

Say you fell asleep in the dark at the end of a long trek, and then woke up in the dawn light to find yourself on the grounds of a palace. The idea that I once actually lived here, among the natural and man-made blessings of Cremorne Point, strikes me now with the same kind of force. This is prime Sydney real estate, with transport connections and wonderful outlooks. I see empty balconies with million-dollar views, and wonder whether the people who can afford to live here ever get a chance to sit and look out over this beauty. It seems an absurd anomaly that there was once a backpackers' hostel five minutes' walk away. Most traveling British kids ended up in the cockroach-infested dives of King's Cross, but for not much more money we got the Harbourside Hotel, a sprawling blue-painted Victorian house backing on to Mosman Bay. At twenty-one, I understood little about how the world worked, and I probably assumed that everyone in Australia lived in a leafy, water-fringed patch of suburban bushland. The realization that I too would be able to live like that for six months seemed a natural right, one of the costless perks of traveling.

It's the hotel I'm looking for now. I set off on the footpath along the east side of the Point, passing a group of middle-aged female ramblers panting up the slope toward me. This was the track we used to follow to the Reserve, the scenic route down to the ferry terminal. The path demarcates the carefully tended gardens of the houses to the left from the equally pristine bushland that drops down through the trees to the bay on the right-hand side. The harbor sparkles between the trunks of eucalypts. The houses look familiar, with their colonial cast-iron balcony railings, but I can't see the one I once lived in. I noticed so little, last time. I have general recognition for the scene, but I can't remember any of its specifics. And that prevents me from picking this grand old house out from that grand old house, and saying: here, this is the place I used to call home.

So far it's not looking good for my experiment in memory. The path becomes darker and more overgrown, and I am farther from the houses. When the inkling of familiarity fades completely, I give up, turn on my heels and retreat along the path to the Reserve. Approached from the other direction, the backs of the houses look even less familiar than before. I get back to the Reserve and join the road that leads up through the spine of the Point on the other side. The winter sun is strong, and I have taken off my fleece for the climb. I feel a weird mix of nervousness, grief and relief, to look and look for it and find that it's not there. If I had found it, I might have been disappointed. But not finding it is worse, somehow. If I'm wrong about this, what else might I be mistaken about?

I didn't expect that this would be easy. I have been in Sydney for five months, and this is the first time I have come back to Cremorne Point to try to find where I used to live. I have wanted to give this journey into my past my full attention. Remembering is an art: it can be done well or badly. I don't want to be distracted when it happens, and I also don't want to catch a chance glimpse of salient details that might contaminate my future reconstructions of the past. I want to be first into the tomb, not the hundredth in a line of explorers, each of whom has changed things in his own way, left traces of his remembering. I am precious about memory; I am aware of its tricks. I actually passed by Cremorne Point once already today, on my way from the ferry terminal at Milson's Point to Circular Quay on the downtown side of the harbor. I averted my eyes when the familiar curve of wooded promontory came into view. Even before I came to Australia this time, I didn't look at the photos from my round-the-world trip. I knew that I would come back here one day, and I wanted the memory to be pure.

The problem is the same as the one I faced on the back streets of Cambridge. If I experience a feeling of familiarity in this landscape, I need to be sure that it's genuine familiarity, dating back to my experiences here all those years ago, rather than the mediation of some intervening, and still very recent, experiences. Already I am risking this, through having to cut back along the path through the Reserve to reach the road. *Of course* it looks familiar; I was here just a moment ago. I am lost in time as well as in space, unable to recognize what is newly familiar from what is more anciently so.

The street address of the hotel had, I thought, long since faded from my memory. The name Cremorne Road sounds familiar, but I cannot retrieve the house number. And yet, the minute I start walking up the waterfront path it comes to me. I lived at number 41. The facts click together, in the same way as the name of an old acquaintance in a photograph pops into your mind as soon as you see his smiling, flash-dazzled face. I start following the numbers of the houses on the right—odd numbers in the twenties, ascending—and realize that 41 is more than just a good guess. The pavement splits away from the road and descends behind a clump of jacarandas screening some new red-tiled houses. At 39, I see a railinged sliproad sloping down from the road above me, petering out into a dead-end turning space. A few steps lead down from it to the pavement. This is where we used to come, laden with shopping and slabs of Tooheys New lager, or where we would stagger out of taxis after nights spent frugally seeing the sights.

But the hotel is gone. The space where the blue house once stood is now filled with a block of cream-colored units in mock classical style. I see a young Asian woman coming out with a bag of rubbish, and I call to her. She looks nervous. I ask if she lives in this incongruous suburban palace behind her. She is the cleaner, she says. I must look suspicious, standing in the road filming the building with my video camera. Saying that I used to live here, in this land of electric garage doors and architect-designed houses, will sound like a feeble pretext. The backpackers would have been flushed out long before this girl started working here, and I'd be better off beating a retreat than arguing for that reality now.

I try the path around the back again. Now that I've situated it on one side, I roughly know, in terms of distance, how far up the promontory the house once stood. But I can't make anything match up from the bushland side. They look like different houses on a different street, as in some optical illusion where the obverse of an object has a completely different physical structure. Have I entered a loop in space-time? Is it the twenty-first century on the road, and still 1989 here? The only thing the two scenes have in common is that the blue hotel is missing from both of them. Then I recognize a bench, planted above a big slab of rock with rain-filled indentations in its surface. I sat out here with my friend Chris one night, our soulful philosophical discussion spoiled only by the pricking of mosquitoes at my ankles. They must have bred in those watery dents in the flat stone. That bit of factual knowledge confirms the recollection. Nothing buzzes there now, but the mosquitoes whine in my memory. And sure enough, a little way farther along is a recently cleared patch where the monstrosity that replaced the hotel now crouches sullenly.

I have not been entirely honest about not coming back to Cremorne Point in the intervening time. I have remembered it, of course, and had conversations with friends who shared the experience with me. But I have also been there in my imagination. Is that the same thing as remembering it? Not quite. I have likened memory to a kind of storytelling, but in my case it's almost true. My memory of this place literally became a work of fiction. In my late twenties I wrote a novel that began with a skinny-dipping scene at a swimming pool not far from my own blue hotel. First novels don't come easily on the whole; they are won at great imaginative cost, particularly their opening scenes. I had set out to get people hooked on the book in their hands, and that had to happen in the first few pages. I set myself to imagine the scene in every detail, and that meant reconstructing the swimming pool and its surroundings entirely from memory. This is the first chance I've had to see whether my fictional rendition is accurate. If anything combines those registers of memory and imagination in unlocking my experience of a place, it will be the bush-lined terraces of MacCallum Pool.

But I don't find it. I follow a cross-route away from Cremorne Road and wander to where I thought the pool lay, only to find nothing but more exclusive condo conversions and manicured lawns. With an anxious, defeated feeling, I walk back along the road in the direction of the ferry terminal. The swimming pool is nowhere. I remember what I wrote about it; I described it as being screened off by rhododendron bushes and accessed from the road by a flight of concrete steps. But there are no bushes, no flights of steps, nothing to identify it. If I am right about having been there, the swimming pool is nowhere near where I remembered it to be.

At the ferry terminal I set off walking along the waterfront path on the west side of the promontory. After a while I see a gate and green metal railings, with a greenish gleam of water beyond. This is the pool that the backpackers of the Harbourside Hotel used to frequent. I sit on a bench on the pavement by the gate and think about how it is possible to get things so wrong. The pool is right down by the harbor, not set up above the road—logical, given that Sydney's municipal pools are filled with harbor water. I realize that I was only able to invent the swimming pool as I did because I had never actually been there. If I had ever come this far along the waterfront path, I would have understood the real location of the pool, and it would have invalidated my hard-won invention.

What I remember is someone else's experience of the pool. Other people came here, and they told me about it. My friend Pam broke her leg here after she was pushed into the pool during a bit of drunken horseplay. But I didn't come here myself—or if I did, I have forgotten the experience so completely that not even the act of reimagining it can spark off any authentic memories. On the other hand, I find it impossible to believe that I spent all this time living at Cremorne Point and never actually went to MacCallum Pool. I am disoriented. I have worked out, after some struggle, where I am in space, but I am lost in my own past. I can say this in all honesty: I really don't know whether I have stood on this spot before or not.

Perhaps this disorientation should not surprise me. I expended so much energy imagining that scene at the swimming pool that it became real to me. I forgot—or perhaps I never really knew in the first place—that I had never actually been there. To imagine something is to park it outside time and space. As we shall see, the scientific study of experiences like mine is full of reminders of this link between fantasy and memory. The products of the imagination take on lives of their own. Without a reality for them to be dragged back to and matched up against, they are free to bloom and flourish in their own way—and even to persuade us that they actually happened.

It is eight years since I went to Cremorne Point to try to find the hotel and the pool. Now, in writing this, I have succumbed to an urge for confirmation. From the jottings in my notebooks I notice that I took some video footage that day. I have transferred the tapes to the computer, but I recall that there was a problem with the transport on some of them, the tapes got chewed up, and a bit of footage was lost. The date of my visit was noted in my notebook, so I can use the digital time stamps in the movie software to find my place. Incredibly, only a week's worth of footage was lost from our entire six-month sojourn in Sydney, and the day of my visit to Cremorne Point falls exactly into the middle of that lost period. I can't go back and look at the footage to confirm how the Point and the pool looked to me, on that day of utter lostness. I have been brought here by imagination, and now I am stuck with the memories.

3

THE SCENT MUSEUM

THE ATTIC OF the house in The Hague was a cramped space under a sloping roof, too small to stand up in. Silvia was well used to the steep spiral stairs that led up to it, as her sister's bedroom had been converted from the attic space and Silvia was a regular visitor. She was home from Leiden University for the weekend, taking a break from her preclinical medical training. She had gone up to retrieve something from one of the boxes that filled the storage space, although she no longer remembers what she was looking for. She was sitting on the floor, rummaging through a box, when a pewtery gleam caught her eye. She pulled out a stainless steel ashtray, with a round shape and a prominent metal button sticking up in the middle. It had belonged to her grandmother Charline, who had died about five years previously. Her *oma*, a writer who had published two books on Greek temple friezes, had smoked cigarillos as she worked and tapped the ash into this ashtray. The cigars were a brand called Meccarillos, hard to find in Holland and so brought back by the family from holidays in Switzerland. The button in the middle of the ashtray spun the tray around, sending the ash to collect in the bottom. As children staying in Charline's big house in Loenen and given the treat of visiting her writer's hut in the garden, Silvia and her siblings had loved pressing the button to make the mechanism spin. When Silvia was ten, Charline had the first of her heart attacks and gave up the cigarillos. She also moved back to The Hague. Silvia has no memory of her smoking at this later stage and, although she does remember seeing the ashtray in her *oma*'s new flat, she would have felt herself too grown-up to play with it.

It was that latter period of her grandmother's life that Silvia was thinking about as she turned the ashtray in her hands. She remembers being surprised that her mother had kept it, as no one else in the family smoked. After quitting the Meccarillos, Charline had lived in sheltered accommodation nearby, and Silvia often went to visit for coffee or dinner, which they sometimes took in the restaurant for residents who did not want to cook. Holding the ashtray, Silvia was thinking about those times and the chats she had had with the old lady as a young girl. The memories were warm, with visual details of the old woman's flat, the sights of the manuscripts and typewriter that the children were not allowed to touch, the pictures of the Greek friezes spread out extravagantly over large surfaces. She had a memory of her *oma*'s voice, the way it sounded in her head, but not the specifics of their conversations.

Then, without thinking, Silvia pressed the button. Suddenly there was the intense smell of the cigar ash, rising from the base of the ashtray. The effect on Silvia was instantaneous. It was a jolt, she remembers; not unpleasant, but not pleasant either. It jarred against the remembering she had been doing, because it wasn't immediately clear what the new memories were about. They seemed to be feelings as much as images. At the same time, she felt very strongly that she *was* remembering. The reminiscences of a moment before, of her *oma* in the serviced flat, were coherent and ordered, neatly arranged within the timescale of her grandmother's life. But the smell of the ash was a new element that didn't quite fit in to the narrative. She had images in her mind from that earlier time in her life, of visiting her grandmother's big house in Loenen: a diamond-shaped window in an internal door, a little grass bridge in the garden. But she also knew that certain of her memories would have been implanted by conversations she had had about them, or by the intervening media of photographs and cine film. When her brother was born, for instance, Silvia and her sister had been left in the big house by themselves while Charline wrote in her hut. They had been given a clock drawn on a piece of paper, and instructions not to disturb the writer until the real clock looked like the depicted time. Silvia was four at the time, and has no direct memories of the event. But she was told about it afterward, and so it has come to have some reality for her.

The feelings triggered by the ash smell were different. From the timing of events in her life, and the fact that her *oma* quit smoking when Silvia was about ten, she knows that her exposure to the cigar smell must date from her first decade. The feeling, deep and emotional, did not last long: it was over in just a few seconds. It was an odd feeling, she says, not linked to any particular memories, but rather a reminder of something that she had completely forgotten she had remembered. If someone had asked her beforehand whether she remembered the smell of the cigars, she would have said no, it was too long ago, and all cigars smell the same. But as soon as she smelled it, she recognized it very distinctly.

Within this emotional onslaught was a strong element of surprise. Silvia had gone up to the attic fully expecting to see things that belonged to her kindly, unconventional *oma*, but smelling them was another matter. The memories themselves were not unpleasant, but the fact that she had been so thoroughly ambushed by them was disconcerting. She remembers going back downstairs and telling her mother about her strange experience, how the forgotten ashtray still had that smell of *oma*. Now in her early forties, she has never sniffed the ashtray since and never smelled that distinctive whiff of cigarillo ash again.

*   *   *

A STORY SIMILAR TO SILVIA'S has become enshrined in one of the most famous passages in modern literature. In Marcel Proust's masterpiece *À la recherche du temps perdu* (translated as *In Search of Lost Time*), we follow the efforts of the narrator (also called Marcel) to reconstruct the story of his childhood and youth. At the beginning of the book, Marcel's memory has failed him. His childhood in the French village of Combray is lost to him, with only fragments remaining. One cold winter's day, when offered some tea by his mother, he tastes a piece of a *petite madeleine* cake steeped in lime-blossom tea, and the effect is immediate but mysterious:

> *No sooner had the warm liquid mixed with the crumbs touched my palate than a shiver ran through me and I stopped, intent upon the extraordinary thing that was happening to me. An exquisite pleasure had invaded my senses, something isolated, detached, with no suggestion of its origin. And at once the vicissitudes of life had become indifferent to me, its disasters innocuous, its brevity illusory—this new sensation having the effect, which love has, of filling me with a precious essence; or rather this essence was not in me, it was me.*

What follows is one of the most famous examples of remembering in literature. Marcel's overwhelming emotional reaction to the taste of the madeleine is not the end of his search for his past, of course. Rather, it is the beginning of an effortful process of reconstruction that stretches out over several more pages, and in which the fragments of his past begin slowly and painstakingly to reassemble themselves. Ultimately, this project of remembering will keep Marcel busy for the three thousand or so remaining pages of the novel. In Silvia's case, the smell of the cigar ash could very quickly be associated with her dead grandmother; she had been thinking about the old lady just moments before. Marcel's reaction is not so easy to interpret. He senses that the "extraordinary thing" is connected to the taste of the tea and the cake, but also "that it infinitely transcended those savors, could not, indeed, be of the same nature." Marcel actually faces quite a struggle to make sense of his feelings at the moment of tasting the madeleine. Further tastings do not work, at least not initially. "It is plain," he writes, "that the truth I am seeking lies not in the cup but in myself." The only way forward lies in deep, repeated plunges of introspection, after which, eventually, something begins to stir: "I can feel it mounting slowly; I can measure the resistance, I can hear the echo of great spaces traversed."

Even this effort is not quite enough, and Marcel has to repeat the examination of his own experience another ten times. It is as though the record of the remembered taste needs to make contact with some other fragment of experience, and they don't quite speak the same language. The Dutch psychologist Douwe Draaisma has observed that the "Proust phenomenon," as it has become known in the psychology of memory, is actually a misrepresentation of what happened to the narrator of Proust's masterpiece. Marcel is not immediately propelled back into the past, to re-experience the remembered moment in a detail at least as vivid as the present one. It actually takes some time for the narrator to link the tastes of the tea and the cake to the memories he is seeking. In that sense, Proust's moment is not a "Proustian moment." The only thing that is immediate is that feeling of "exquisite pleasure"; everything else is still some way in the distance.

When autobiographical memories are triggered by sensory experiences, a process of meaning-making often needs to occur. Sensory impressions at the point of recall can become invested with memory meanings, as in Marcel's case, through a process of active reconstruction. Later, when the stimulus has become associated with the memory more than once, the encounter with the sensory impression can trigger the meaning immediately. When I asked around for anecdotal examples of Proustian memory, I heard several stories in which smell had a power to trigger immediately interpretable memories. One friend told me that the smell of fresh basil reminded her of the first weeks she had spent living together with her husband, as she had bought a big basil plant for the windowsill of their new kitchen and the smell was at that time still unfamiliar to her. Another friend remembered how the scent of cloves in a pomander was immediately redolent of her grandmother's house: the ticking of the clock, the coal fire, the cramped kitchen, the toffees in a glass bowl on the sideboard because her grandfather had given up smoking.

In the case of Proust's narrator, the sensory impression is not immediately eloquent by itself. It is an indecipherable message from a past time, and Marcel is painfully aware of the effort he must make to bring the associated memories into consciousness. "What an abyss of uncertainty, whenever the mind feels overtaken by itself; when it, the seeker, is at the same time the dark region through which it must go seeking . . . It is face to face with something which does not yet exist, which it alone can make actual, which it alone can bring into the light of day." In this passage, which is so frequently cited (but perhaps less often properly examined), Proust shows himself to be fully cognizant of the reconstructive nature of memory. An autobiographical memory must be built up from the available materials, and a telling sensory impression is just one of the components of that edifice.

Painstaking as this process proves to be, however, Marcel's final revelation is sudden. Just as he is giving up hope, he realizes that the taste that has so assailed him is the taste of a comparable piece of madeleine that his aunt regularly used to give him (after first dipping it in her tea) on Sunday mornings at Combray. As soon as that connection is made, the remainder of the context springs up instantaneously. The effort is in the initial task of connecting the taste of the madeleine to the memory of Aunt Léonie. When the image is retrieved, other images and meanings snap into place. Proust's own interpretation was that smell and taste have a particular power in reawakening buried memories, but also that they present a specific challenge in their inability to communicate in the language of other memories. Proust's formulation of the persistence of such memories is itself utterly memorable: "But when from a long-distant past nothing subsists, after the people are dead, after the things are broken and scattered, taste and smell alone, more fragile but more enduring . . . remain poised a long time . . . amid the ruins of all the rest; and bear unflinchingly, in the tiny and almost impalpable drop of their essence, the vast structure of recollection."

Was Proust right to say that smell and taste have these special powers? These senses certainly have some odd neurological properties. In most of our five senses, the information received by the sensory organs is transmitted to the memory systems of the brain via the waystation of the thalamus (a part of the brain just above the brain stem and enclosed by the hippocampus). In the case of smell, evolution has arranged it so that the olfactory receptors high up inside the nose send their signals along a short pathway to the olfactory cortex and then straight to the hippocampus, bypassing the thalamus. This is often taken to explain the special power of smell in memory. And if smell works in that way, then taste will follow. We can actually only detect a very small range of basic tastes; the complexity of the tastes we experience is largely due to the functioning of the olfactory, or smell, system. When Proust experienced the special, redolent savor of the madeleine, he was smelling it as much as he was tasting it.

But neuroanatomy is only part of the picture. Concluding that one sense has a special power in memory just because its pathways to the brain are shorter and more direct than those of our other senses is likely to lead us astray. In the visual system, the pathways from the light-receptive cells in the retinas to the visual cortex are long and convoluted, and yet no one would deny how important the sense of sight is for autobiographical memory. Indeed, Proust is arguably a much more visual writer than an olfactory one. As any novice wine-taster will be able to tell you, trying to find words for aromas is difficult, and so perhaps it's no surprise that a writer like Proust would fill his book of memory with visual impressions.

In order to conclude that smell has a special role in memory, we would have to show that olfactory memories are more deeply rooted than other kinds. One line of evidence indicates that smell-triggered memories stretch particularly far back in time. A British study showed that when autobiographical memories were cued verbally, there was a predictable peak between the ages of eleven and twenty-five: the well-replicated phenomenon known as the reminiscence bump. When smell cues were used, in contrast, the peak occurred earlier, between the ages of six and ten (the period of time tagged by Silvia's ash smell). Cueing memories with smell seems to shift the reminiscence bump back toward earlier memories.

Why should smell memories be older? One clue lies in the fact that smell sensations are so hard to put into words. As we shall see, talking about the past is important for children's development of autobiographical memories. Smell memories may have a particular reach back into the past because they stem from a time before memories were encoded in language. This fits with the idea that the Proust phenomenon is initially a rush of emotion, hard to put into words. The writer Richard Holmes points out that Proust's observation was predated by a certain keen-smelling mammal, several years before Proust's work was published.

> *It was one of those mysterious fairy calls from out of the void that suddenly reached Mole in the darkness, making him tingle through and through with its very familiar appeal, while as yet he could not clearly remember what it was. He stopped dead in his tracks, his nose searching hither and thither in its efforts to recover the fine filament, the telegraphic current, that had so strongly moved him.*

Mole's struggles to identify the distinctive smell of home in Kenneth Grahame's children's classic, *The Wind in the Willows*, point to another aspect in which smell memory is different from other kinds of memory. The psychologist Rachel Herz has spent much of her career investigating the emotional power of smell-cued memories. In one recent study, Herz presented participants with three items that could potentially cue memory: popcorn, freshly cut grass and a camp fire. In order to provide a careful comparison across sensory modalities, Herz used three different ways of presenting each of these items: as a short movie clip (visual), as a sound (a five-second sound clip) and as a smell. Participants were asked to generate autobiographical memories in response to these cues, and then to rate their memories on scales such as emotionality, vividness and specificity. The smell-cued memories turned out to be more emotional and evocative than those recalled to visual or auditory cues. They were not, however, more vivid or specific. In a separate study, Herz and colleagues have shown greater activation in the amygdala and hippocampal regions for personally significant memories that are odor-cued, which fits with the neuroanatomical evidence that the smell system has a direct line in to the memory and emotion networks of the limbic system.

Not all of these findings on the specialness of odor-cued memory have been reliably replicated by other studies, and there is still some debate about the extent to which smell memory can really be considered distinctive. Smell-cued memories may be more emotional, but they are not more accurate, vivid or specific than other kinds. They typically date back to earlier life stages than memories cued in other ways, shifting the position of the reminiscence bump to childhood instead of early adulthood. The effectiveness of odors as a cue may not be limited to past events, either. In one recent study, university students were presented with smells such as mint and whisky and were asked to imagine events that might happen to them in the future as well as recall those that had occurred in the past. The results replicated the backward shifting of the reminiscence bump for smell memories, with most such memories tending to be from the first decade of life. They also showed that smells can act as effective cues for the imagining of future events, but only for events imagined happening in the next year or so. As we shall see, the fact that cues that unlock the past can also unlock the future points to one of the most interesting new areas for memory research.

One thing that is certain is that Proust did not break new ground in proposing a special link between smell and memory. The writer Avery Gilbert has pointed out that the neuroanatomical connection between nose and brain had been accurately prefigured by the physician and writer Oliver Wendell Holmes in 1858. Countless other writers in France and the United States had cited the special mnemonic power of smell long before Proust began his novel. An otherwise obscure clinician, Dr. Paul Sollier, who treated Proust for neuraesthenia shortly before he started work on the composition of *À la recherche*, had developed his own therapeutic program based on the power of involuntary memory, and treated Proust with it. Other unacknowledged debts in Proust's work are to the philosopher Henri Bergson, whose theory of "pure" memory—dynamic, episodic and largely involuntary—was published in 1896, and to the works of the evolutionary biologist Richard Semon and the psychologist and philosopher William James. Among writers, one of the most vocal champions of the power of smell was Rudyard Kipling. In his poem "Lichtenberg," the narrative of an Australian soldier far from home in the Boer War, the smell of a wattle is instantly redolent of the similar plants of his homestead in New South Wales:

> *Smells are surer than sounds or sights*
> *To make your heart-strings crack—*
> *They start those awful voices o' nights*
> *That whisper, "Old man, come back!"*

* * *

MOST OF US ARE CONSCIOUS of the power of smell to evoke memories, but few of us try to put it to a positive use. One exception was the artist Andy Warhol. A self-confessed perfume addict, Warhol would deliberately wear the same cologne for three months at a time, until its smell was associated with his memories of that period. Then, long before he had tired of its scent, he would force the perfume into retirement and never go back to wearing it again. This olfactory ruthlessness led him to build up an extensive collection of partly used bottles of scent. "Of the five senses," he wrote, "smell has the closest thing to the full power of the past." If Warhol wanted to go back to a particular point in his life, he would drop into his scent museum, unstopper the relevant bottle and take a whiff. Seized by a mood to re-experience certain memories, he would use the power of scent to transport himself. It was only a temporary measure. As in Silvia's case, the experience would last only for as long as the sensory experience did. When the smell faded, so did the associated memory.

With characteristic irony, Warhol claimed to appreciate the neatness of this tool for reminiscing, and the technique's pleasant lack of after-effects. Warhol's story is interesting for all sorts of reasons, but one is the way in which he very deliberately set out to harness the mnemonic power of smell. In contrast, Marcel's moment with the madeleine is usually held up as an example of the power of *involuntary* memory: the past's sudden, unexpected assault on our consciousness. Warhol realized that the Proust phenomenon did not have to be a random, involuntary experience (as his own earlier adventures in smell memory had been), but rather that it could be harnessed. Warhol does not really explain what it meant for him personally to build up a stock of smell memories; in fact, his account seems more interested in his scent museum's value as a cultural repository of vanished smells than as a treasure house of personal memories. If Warhol really did use his vast collection of partly used cologne bottles as a museum of his own past, he keeps quiet about the reminiscences that resulted. He prefers instead to itemize the ambient smells of New York City: the hot dogs and sauerkraut carts, the Moroccan-tanned leather on the street racks, the mildewy smell of Chinese import stores.

Most of us, I would guess, do not go to the same lengths as the legendary pop artist to keep our noses in touch with the past, but we may do it to a lesser degree. One friend told me that she hoarded a jar of Boots Country Born hair gel whose scent reminded her of getting ready, as a teenager, to go out nightclubbing in her Lancashire hometown in the early 1980s. Another friend still dabs on Chanel Coco, her first-ever perfume, to remind her of her youth, while a third will sometimes crush a mint leaf between her fingers to remind her of her grandmother's summer garden. One respondent to a classic study of smell memory published in the 1930s said that he used a bunch of sage brush to remind him of the Nevada desert where he grew up, finding that "a slight sniff doubles and redoubles that tranquil nostalgia." Some psychologists will even recommend the use of smell as an exam revision aid. Rachel Herz suggests that those studying for exams allow a powerful smell, like a camphor lip balm, to become associated with the long hours of revision. Reapplied (quite legitimately) in the exam hall, the smell ought to be effective in reigniting memories of what has been learned.

All of these techniques are, at some level, about exerting control over something that is usually understood to be automatic. Even in Marcel's case, with his intense intellectual effort to interpret his fleeting madeleine feeling, we see the voluntary taking over from the involuntary. But Proust also recognizes that too great an effort to remember can be counterproductive, and that sometimes your best chance of remembering is to allow yourself to forget. This is his strategy in the scene with the madeleine, where he forces himself to clear his mind and think about something else after his first unsuccessful attempts to harness the memory. The literary critic Roger Shattuck points out several other occasions in which Marcel exploits his own forgetfulness in order to go further with memory than most narrators can. In each case, voluntary efforts at recollection have only limited power, and the triggering moments are unexpected and surprising. Involuntary memory, as Proust termed it, follows its own laws, and does not submit readily to the controlling hand of the human will.

Some have even gone so far as to wonder whether involuntary memory constitutes a special category of memory, with its own rules and neurological systems. One view is that involuntary memories are particularly likely to be triggered by sensory cues, as in the classic Proust phenomenon. This idea was examined in a study in which participants were asked to record their involuntary memories along with the cues that triggered them. The researchers found that involuntary memories were more likely to be triggered by abstract cues (such as thoughts) than by sensory ones (such as smells and tastes), concluding that involuntary memory retrieval was not all that different from its voluntary counterpart. In contrast, other studies have shown that involuntary memories have richer episodic features than voluntary ones. For example, some Danish researchers asked their participants to record in a notebook details of both involuntary and voluntary (word-cued) memories. They found, in support of the Proustian view, that involuntary memories were more focused on specific episodes, and also had a greater emotional quality.

Andy Warhol's story also emphasizes how smell memories are susceptible to interference. In the early days of research into the Proust phenomenon, it was claimed that smell memory is particularly resistant to forgetting. But more recent studies have shown that the same interference effects apply to smell memory as to memories cued in other modalities. Warhol recognized this, as evidenced by his retiring his scents after a certain amount of active service. By making sure that he never wore a scent again, he could prevent other memories from becoming entangled with it. He writes of his concern that there might not be enough scents in existence to allow the experiment to continue indefinitely, but reassures himself by recalling his visits to the perfumeries of Europe and the vast array of fragrances on sale there. He need not have worried. Thanks to the work of Nobel Laureates Richard Axel and Linda Buck, we know that the human olfactory system can encode around ten thousand different smell patterns, due to the concerted operation of around four hundred different olfactory genes, each of which controls the production of a single specialized protein receptor.

Most of us cannot afford to invest in that range of perfumes, however, and so smell memories are susceptible to being overwritten by subsequent fragrant episodes. In that smell survey from the 1930s, one respondent noted the power of the earliest associations to win out, eventually, over later intervening ones. In his case, the smell of a woolen overcoat was associated with memories of his uncle, but later became "reconditioned" to new connections. Ultimately, the original association reasserted itself. Perhaps, the respondent suggested, numerous new memories become associated with the smell in question (especially if it is a frequently encountered one), and end up canceling each other out, leaving the original memory as the sole survivor.

The question of the persistence of early smell memory associations was directly addressed in a recent study conducted in Israel. Noting the particular strength of the earliest associations made between a smell and an object, the researchers presented participants with objects associated with different pleasant and unpleasant odors (such as pear or fungus), followed about an hour and a half later with a different pairing. To determine whether there was anything special about smell memory, they also tested memory for associations between objects and sounds (such as a guitar or an electric drill). A week later, memory for the associations was tested. Participants saw objects they had previously seen and were asked to recall which odor or sound they associated with them. The results showed that objects were associated more strongly with the smell or sound they had been paired with first in the lab session the previous week, but that that pattern held only when the smells or sounds were unpleasant.

So far the data seemed to support the idea that first memory associations are privileged (at least for unpleasant stimuli), but did not point to any difference between smell and sound. But the researchers were also scanning their participants' brains during the memory tests. Here, the findings revealed a difference between the sensory modalities in the activity of the left hippocampus. For smells, there was more activity in this region for the first association compared with the second, whether the smell was pleasant or unpleasant. So strong was the pattern of brain activity that the researchers could predict, just by looking at the fMRI scans from the first testing session, which smells would be associated with the object a week later. The Israeli research seems to suggest that the crucial issue is how a smell is dealt with the first time it is encountered in a particular context. Subsequent interference may play a part in our forgetting certain associations, but the distinctive power of early smell memories remains.

It seems likely that smell-related memories can be so deeply ingrained at least in part because the smells of our childhoods do not recur frequently enough for them to form new associations. One friend told me that the unlikely combination of smells of fish and chips and fresh paint will be forever connected with her memories (from around the age of six) of helping her dad to paint the cricket clubhouse. Smells form a part of our environment that we don't always attend to (not least because they are so difficult to localize in space), and it may be that fact that partly accounts for their power. As Avery Gilbert notes, smell memories surprise us because we weren't attending particularly to the smells when the event happened, but we made the connections anyway. "Because odor memories accumulate automatically, outside of awareness," he writes, "they cover their own tracks. We don't remember remembering them."

But smell is not the only sense that can ambush us in this way. Music is another invisible sensory stimulus to which we don't always pay full attention. Any record or CD collection is a kind of sound diary analogous to Warhol's scent museum. When songs are remembered, they are encoded along with a whole variety of contextual details, and subsequent re-exposure to the song can bring some of those details back into consciousness. One friend had a vertiginous attack of nostalgia recently when she listened to an Internet radio station playing hits from the 1980s. They were not favorite songs of hers, so she had not heard them much since that time, and it was that immunity from subsequent interference that made them particularly strong cues to memory.

A recent neuroimaging study capitalized on this effect by eliciting autobiographical memories in a sample of sixteen young adults. The bits of music used were thirty-second clips from popular songs taken from the preceding ten-year period. As the participants listened to the clips in the scanner, they were asked to think silently about the memories that came to mind, and then indicate how specific each memory was (they pressed one button if the memory corresponded to a general period in their lives, another button for a memory of a general event, and a third for a specific event memory). They also rated the memories for emotionality, vividness and several other factors. The results showed that the musical cues triggered highly emotional memories that were reported as not having been retrieved very much previously. As expected, the autobiographical memory system of the medial temporal lobe was thoroughly activated, as well as regions in the ventromedial prefrontal cortex (deep down in the front of the brain) and posterior cingulate (in the middle toward the back). For more specific events, the dorsomedial prefrontal cortex was particularly busy, fitting with the idea that this area of the brain is particularly important for memory searches for specific events. Specific events were also retrieved significantly more quickly than general events and lifetime-period memories, which suggests that some kind of Proustian "direct retrieval" mechanism was at work.

Music-cued autobiographical memory can also demonstrate the power of first associations. A song that might have been heard many hundreds of times can nevertheless send the listener back in time to its first hearing. The Israeli smell-object pairing study did not show any specific neural patterning for auditory stimuli, but the researchers only studied single sounds rather than musical phrases. Their behavioral data, however, did show that the "first is strongest" principle applies to auditory stimuli. And precisely the same principle may apply to stimuli in other modalities. When I light the stove in the study in which I write, I have a recurrent memory of a time a year or so ago when a bird was trapped in the chimney. I have lit the stove scores of times since then, and yet the memory recurs. Certain memories seem to get burned in, even when they do not coincide with the first exposure to the sensory stimulus, and when there have been many subsequent exposures that might potentially have caused interference. One fact that distinguishes the memory I associate with the stove is that it was emotionally powerful. It was very unpleasant and distressing to hear a bird trapped in the chimney, knowing that I could do nothing to rescue it. As we will see, it may be that emotion is a particularly important factor in determining which memories become burned in and which are left to fade.

Our different senses are likely to interact with memory in distinctive ways. To my mind, the most interesting question is not whether there is anything special about smell memory or musical memory, but how any sensory stimulation can lead to the recall of autobiographical memories. Olfactory memories may be particularly difficult to integrate because of their distinctive neural pathway, meaning that the initial reaction is more emotional, and because they are somewhat refractory to language. But the fact is that the brain can somehow integrate sensory information from different channels into vivid, multidimensional autobiographical memories. Studies have shown that when a smell, for example, is paired with a picture, subsequent presentation of the picture can spark activation in the smell areas of the cortex. The brain remembers sensory associations through seamless collaborations between the medial temporal lobe system and the different parts of the cortex that process the information received by our five senses. By emphasizing the multimedia quality of remembering, these examples of fragrant and songful memories show how closely the making of autobiographical memories is linked to our sensory and emotional experience of the world.

4

THE SUNNY NEVER-NEVER

I HAVE PUT the envelope someplace safe, waiting for a quiet moment to go through it. My sister has written my name across it in blue Biro, marking the beige of the manila with a few letters of her cheerful handwriting. We have been clearing out our aunt's house, still slowly picking through boxes of hoarded treasure more than a year after she died. As Clare, my sister, has the shorter journey to the house in Gloucestershire, she has been continuing the sorting on her own, and she has saved this envelope of relics for me. Every year we wrote thank-you letters and sent copies of our school reports, and Auntie Sheila kept them all. From the thickness of its contents, I can tell that the archive is not large, but still I want to wait until I have the house to myself before I go through it. As in my coy approach to the ferry terminal at Cremorne Point, I have averted my eyes, conscious of how my memories may be contaminated even as they are taking shape. Remembering is a serious business. It demands attention. For a journey into the past, you have to pick your moment.

I open the envelope. Inside are pieces of paper of different sizes and colors, covered with childish handwriting and drawings. The characters my younger selves formed are made up of simple shapes, squares jammed against circles, with no attempt to join them up into cursive words. I read each carefully preserved document and eavesdrop on the life it depicts. I have seen none of these letters since I wrote them, and I'm keen to remake the acquaintance of the person, or people, I was then. As a five-year-old, I am excited about a forthcoming trip to visit my cousins in the Caribbean, in what will be my first flight on an airplane. There is a letter I wrote at age six, thanking my aunt and uncle for the dinosaur book they gave me for Christmas. At eight, I am obsessed with bird-watching, and I am keen to share every ornithological detail with Sheila, herself a keen naturalist. The words themselves—my childish efforts to portray a life—are not particularly strong cues to memory; they are too full of stock well-wishings and formulaic thank-yous. But certain details trigger warm, uneasy feelings of familiarity. It is long since lost, but I *remember* that book about Greece and the serpent's teeth, gifted to me on my fifth birthday. Those notelets with the yellow frogs sitting spottily on lily pads: they're known to me. And yet there are details I surely ought to remember whose telling leaves me cold. I don't recall, for instance, my parents coming into my classroom and congratulating the five-year-old me on coming top of the class. I don't remember my parents being together, except in one or two fragile images. When I do focus on real details, they have a bizarre, random specificity: the color of the linoleum in the house we moved to after Mum and Dad separated; the sign announcing the show house in the new estate we moved to. There is no mention of the divorce itself, or the emotional upheaval that must have accompanied it. I'm too eager to set my aunt straight about the decor, the bird sightings, the nitty-gritty of this ripped-apart life. These are missives from another existence, a way of attending to the world that is alien to me. If I was ever that person, I have forgotten what it was like to be him. His letters don't bring it back.

Why do I care? Why have I protected this moment so assiduously?

Because, like every human being, I want to know who I am, and part of that is knowing where I have come from. I'm not looking for anything specific, some veiled trauma or epiphany that set me on this path as a person. I simply want to connect, to see continuity. There were indeed life-changing events, moments that from this perspective seem to define entire epochs in my childhood, but my letters are silent about them. I account for myself no differently before and after. And I know I can rely on the honesty of these cues. The letters before me have not been talked about in the intervening years. They have lain in a drawer in my aunt's labyrinthine filing system; it took her death to release them back into the world. There has been no mediation by intervening acts of remembering. It's not often that we walk open-eyed into such a pure encounter with our pasts. If I was ever to have had a chance to find out how I became who I am, it would have happened today.

Some of the details may be strange, but there is an irrefutable logic to the idea that I am the same person as the boy who wrote those thank-you letters. The cells in his body may have been replaced several times over, but I share a name with him, and a history. When he looked out at the world, he registered different details from the ones that would inter-est me now, and his spelling was rather bad, but we would have shared many features of our personality. How far back would I have to go before I started to doubt that continuity? Further back than the letter-writing evidence would stretch, for sure. With my very earliest memories I am on my own, with no souvenirs or aide-mémoire to help me out. What about if I cast back, unaided, and ask myself: What's the first thing I remember? If I do that, images come to me. I see a toddler pushing a forklift truck along a pale carpet. The room is thick with late-summer sunlight. There is no sound. If it's me in the image, I have not yet reached the age of three years and four months. I know that for sure because I moved house, with my parents, at that age; the house we moved to became my most familiar childhood home, and this memory image does not fit its architecture. This must have been the beige carpet in the living room at Harrow Way, the house where I was born. It's my only memory of that place. Not even the photos I have seen of it have infected my memories, trying to trick me into believing that I remember more of it than I do.

Here's another, later memory. I am standing at the top of the stairs in the new house, and I have just shat my pants. I am being called downstairs, probably for tea, and for obvious reasons I don't want to go. I am three years old. I know this because the stairs are those of the new house in Danbury, where we moved in the summer of 1971. (My mother confirms the date of the move. Memory can be a science, too.) But I know this fact from the inside as well. I remember thinking: *You are three, you are long out of nappies, you should not be doing this*. I'm not having this explicit, logically detailed, grown-up thought, of course: I am simply feeling the shame and wrongness of it. I can remember the blue briefs I was wearing. I am absolutely sure about the light from the lamp that hung over the banister at the foot of the stairs, the airless fustiness of the hallway. I relive the moment, and I think I relive it in pretty much the same way as I relived it the last time I remembered it. I would resist anyone's efforts to deny it, to tell me that it was a fiction, to undermine its reliability. I would resist it because it is a part of me.

This is not a new encounter with my past, like my reading of Sheila's archive of thank-you letters. I have asked myself about the dawn of memory before; others have asked it of me, too. Recalling your earliest memory is a familiar parlor game, a popular ritual for cherishing something that we, as a species, value tremendously. Look, we tell each other, this grainy, sun-drenched snapshot is who I am. And because I have rehearsed this self-definition many times before, I have to be much more on my guard against false remembering. The carpet memory is particularly unreliable. It barely counts as a memory; it's just an image that comes into my head when I ask myself the question. I don't really know where it comes from. I also don't see how it is logically plausible that I could have it, as I figure objectively in it. I am the toddler in the picture. On the day in question, I would have been down there on the carpet, looking out at the room through my own eyes. So why am I the actor in the drama? Why am I not seeing it from my own point of view?

Actually, this turns out to be a common feature of early memories. Memories in which the rememberer appears as an objective, third-person character are known as *observer* memories, and are distinguished from *field* memories, in which the memory preserves the original field of view of the character doing the remembering. The logical implausibility of observer memories was of interest to Sigmund Freud, who saw them as evidence that early memories are reconstructions of the episodes on which they are based. The shift of perspective from first person to third person indicates that some kind of fiction-making has been at work, and that the memory cannot be a true representation of that moment of experience. If point of view has slipped, what else might have been tampered with? The fact that early memories can be so banal was another cause for suspicion. Freud argued that these bland, third-person memories acted as screens for other, more significant events that our developing egos have repressed.

The memory of soiling myself on the stairs is different. Here, I am at the center of the memory. I am reliving the event: seeing the same things, feeling the same feelings. It is a field memory, not an observer memory. I know where it comes from, unlike the carpet image. I am also much more certain that it is trustworthy. For understandable reasons, my mishap was an event that was never discussed at home. Mum probably found out about it afterward, but she never mentioned it. Or perhaps she never even knew about my accident. I remember that it was a hard little turd, easy to dispose of and leaving not much of a mark. The fact that I could deal with its consequences myself meant that it could stay private, and so I can't be remembering someone else's take on the occasion.

Why do I remember these events and not others? Why aren't my memories full of the big stuff: my parents in the classroom, congratulating me on coming top of the class; the inevitable tensions that would have prefigured their separation? Memory is constantly surprising us, and sometimes infuriating us, with its randomness. In the Dutch writer Cees Nooteboom's phrase, it is "a dog that lies down where it pleases." I think I remember the stairs occasion because it was humiliating. We have particularly strong memories for early insults to our egos. But the carpet memory seems completely random. Memory has chosen to settle where it likes, and it has lain down here. For some reason, something stuck in my mind that day. I will never know exactly why, and that intriguing randomness is one reason I will never completely be able to trust the recollection.

Bland or momentous, the striking thing is how little remains from that time. These two memories, different as they are, account for everything I can remember with any certainty from before the age of about four. And I don't think I'm unusual in falling victim to what Freud called "the remarkable amnesia of childhood." When you ask people to reminisce about their earliest lives, they rarely recall events from much before this age. Those memories that do survive are fragmentary, shot through with vivid sensory impressions but lacking the detail and organization of memories from later in childhood. They do not hang together like stories about the past should. "Here and there," Freud wrote, "we come upon people who can boast of a continuous memory from the first beginnings to the present day; but the other alternative, of gaps in the memory, is by far the more frequent."

Freud's observation has been very thoroughly supported by psychological research. In one recent study, researchers in Nova Scotia asked a sample of university students about their earliest memories. After writing down details of the event, the participants had to indicate what age they were when they estimated the event to have happened. The results showed that the average age of earliest memories was about four and a quarter years. Very few memories dated from before the age of about two and a half. Other studies have confirmed this two-and-a-half-year cutoff for memories of childhood. When we ask adults about their childhood, events from before that time are rarely remembered.

What could be the explanation for this childhood amnesia? One simple answer is that our early memories are lost to us because memory itself is a capacity that takes time to develop. But a little reflection shows that this explanation does not work. Infants remember things; they would not be able to learn vital lessons about the world if they could not rely on memory. Using a technique known as *operant conditioning*, the psychologist Carolyn Rovee-Collier has demonstrated impressive feats of remembering among babies before the age of six months. Operant conditioning is the kind of learning where animals, including humans, remember a connection between their own action and a particular outcome: for example, learning that pressing a lever leads to the delivery of a reward. In Rovee-Collier's experiments, infants are placed in their own cots with a soft ribbon attached to one ankle. When they kick their legs, a mobile above them moves. Babies enjoy the movement of the mobile, and will work for this reward: the harder they kick, the more the mobile jigs about. The aim of this kind of experiment is to see whether infants will learn the association between their kicking and this particular mobile's movement. If they kick more than they would do when presented with a different, unconnected mobile, it is assumed that they have recalled the connection between their kicking and the movement of the mobile. Studies like this have shown that two-month-old babies can remember this association for three days or so. At six months, the retention interval is more like two weeks.

I saw plenty of evidence for early remembering with my own children. At ten weeks, for example, my daughter, Athena, demonstrated a modest grip on her past. Under our bathroom sink there used to hang a blue floral-patterned curtain. When we'd finished bathing her, we would lie her in her toweling wrap on her changing mat in the narrow gap between the sink and the door. In those days she would stay where you put her, pinned down by her own weight and weakness, a little upturned frog, all belly and minuscule strugglings. Immediately she would turn to the curtain and look at it, drawn to its big colors and shapes. We soon realized that she was anticipating this visual treat, turning to face the curtain before we had even put her down. She would put a hand out and bat at it, trying out the connection between hand and eye. It became a game for her, a routine to round off bath times. She was remembering the curtain, anticipating its presence as a feature of her increasingly ordered world.

So the reason we forget our early childhoods cannot be because a baby's brain is too immature to do any remembering. The systems may be rudimentary, and due to develop further in important ways, but the structures of memory are in place from our earliest days. There is also plenty of evidence that fetuses can learn while still in the womb, and retain the information through birth and beyond. It's not even correct to say that there are distinct memory systems, one that serves infancy and another that takes over later. When scientists study features such as rates of forgetting and factors that interfere with memory retention, they see that remembering and forgetting in babyhood is too much like memory in adulthood for this to be a convincing explanation of childhood amnesia. A different kind of explanation is needed, one that goes beyond the nuts and bolts of information storage and centers on some big questions about language, identity and consciousness.

The current consensus among memory researchers is that we need other capacities to be in place, skills that are not directly to do with the storage of information, before we can hope to carry our memories forward into later childhood and adulthood. One such factor is language. As soon as you can use words to describe your experience, you begin to have an entirely new way of encoding, organizing and retrieving information about the past. In one recent study conducted at the University of Leeds, the psychologists Catriona Morrison and Martin Conway asked adults to generate childhood memories in response to cue words naming everyday objects, locations, activities and emotions. By looking at existing data on the average age in infancy when these words are acquired, they were able to show that the earliest memories always lagged behind (by several months) the age at which the corresponding word was learned. "You have to have a word in your vocabulary," Morrison observed, "before you're able to set down memories for that concept." As has been noted many times, it is unlikely to be a coincidence that the end of childhood amnesia corresponds to the period in which small children become thoroughly verbal beings.

Another capacity that is thought to be critical has to do with the personal, subjective quality of autobiographical memory. The sort of memory we are interested in is not just about remembering information or associations, but rather of re-creating an image of the world of which we ourselves are a part. This quality of memory is the thing that memory researchers call *autonoetic consciousness*. Autonoetic consciousness is the quality that puts you at the center of your memories, so that you can relive the moments from the inside. It's what makes your memories feel as though they are your own. To a large extent, it is what makes your life feel as though it happened to you, rather than to someone else. The writer Vladimir Nabokov noted that we are always at home in our pasts. We never feel like we are strangers, or we would not feel as though we were remembering. When we visit a past time, we usually look out at its events with our own eyes. The memories feel like they are ours, because we are at the center of them.

The importance of this quality of memory is illustrated by my two childhood memories. In one of them I have autonoetic consciousness; in the other I do not. The stairs memory is a first-person narrative. In the carpet memory, by contrast, I am a character in a drama that I am viewing. My ability to relive my moment of shame on the stairs is the main reason I am more confident in this memory than the other one, and also why I would pin it to a later point in time.

This makes me wonder: was I actually conscious of those two events in a different way? Perhaps the carpet memory is genuine, but I wasn't aware of myself in the way I was to be later, and so the memory lacks that autonoetic quality. Precisely this explanation has been put forward to account for childhood amnesia. The legendary English musican Robert Fripp recalled as his first memory a moment of waking up in his own body, as a baby in a pram, aged between three and six months. Although Fripp's first memory is unusually early, many otherwise comparable early memories come with a flash of realization about the self. Writers such as Nabokov, Edith Wharton, Anthony Powell and A. S. Byatt have linked their earliest memories to an insight into their own identity. To have autobiographical memory, you arguably need some kind of understanding of the self to whom the events are happening.

Perhaps the key to understanding childhood amnesia is not to ask adults what they have forgotten, but to ask children what they can still remember, and how they remember it. It is possible that children will remember things that their later adult selves will not. And if we look carefully for a change of quality in the memories—the point at which they become autonoetic—we may begin to understand the processes that veil our early childhoods from our later selves.

When do young children begin to star in their memories in this way? You certainly get an odd sense, in talking to toddlers about what they remember, that the past is something that happened to other people. Small children can often be quite accurate about recalling things that happened to them, but they do not always seem to recognize themselves as the authors of the memory. Research has shown that many of the memories that young children generate are actually parrotings of things their parents said as the events unfolded. It is as though the child is recalling the adult's description of the event, rather than claiming it as her own. As we shall see, findings like these have founded an entire theory of how autobiographical memory is constructed through social interaction. But they leave us with the problem of trying to work out how and when autonoetic consciousness gets started. If we want to tease apart those memories that children actually re-experience from the ones that they just knew happened, we might have to find different ways of asking the question.

One problem that children face is keeping track of where information comes from. Small children are not particularly good at making judgments about how they know what they know. In childhood, as in adulthood, memory is as much about rerembering an existing memory as it is about finding a direct line to the past. And those existing memories are themselves contaminated by other kinds of information, such as oft-repeated family stories and narrative details picked up from elsewhere. Leonard Woolf, husband of the writer Virginia Woolf, wrote on this point: "Some of the things which one seems to remember from far, far back in infancy are not, I think, really remembered; they are family tales told so often about one that eventually one has the illusion of remembering them."

Because of their relatively weak capacities to monitor the sources of information, children are particularly susceptible to this kind of interference. Memories that are not actually theirs become incorporated into their own developing autobiographical accounts. Even a toddler will have been exposed to other people's tellings of the memory, accounts of similar (and potentially interfering) happenings, and—more and more commonly—digital representations of the events, such as photographs and video footage. This makes it particularly difficult for them to know the difference between remembering an event for real and remembering their experience of some representation of it.

When we ask children about their pasts, we are not simply asking them to remember facts about their lives. We are asking them to travel back in time and reinhabit those moments; in other words, to have autonoetic memories of them. If I ask you about your first day at school, I don't want you to tell me details of semantic memory such that it happened on September 5, 1974; I want you to travel back to the moment and look out at it from within. I want you to tell me what it was *like*, and that means somehow re-experiencing the consciousness of that moment.

Time travel sounds like a far-fetched idea, and literally speaking it is: no one has yet built a machine that would take us back to the future. But psychologists have found "mental time travel" to be a useful concept for thinking about the shuttling between present and past that is involved in memory. If a child is to be able to do autobiographical memory, she must become a time-traveler. She must learn to program her TARDIS for different points in time, albeit on the small timescale of her own life span rather than that of galactic history. She needs to stop being the person she is now, and start being the person she was back then.

This is a formidable cognitive challenge. In a recent article examining the idea of mental time travel, Thomas Suddendorf and Michael Corballis introduced a theatrical metaphor. To do mental time travel, you need a stage: a representational space where you can depict the events of the memory. You need a playwright, some cognitive system or set of rules that can generate an unlimited set of past and future scenarios. You need actors, which requires an understanding of characters—their thoughts, feelings, intentions, beliefs, desires, and so on—and the interactions among them. You need a theatrical set, which relates to a basic understanding of the physics of how time and space work. You need a director, capable of evaluating different possibilities and choosing between different possible representations. You need an executive producer, who can give you some voluntary control over how the representation is unfolding. And you need a broadcaster: some way of telling the world about the memory you are having, which usually means a way of encoding it into language.

Each of these capacities is an area of interest for developmental psychologists in its own right, and each probably requires the orchestration of a suite of sophisticated cognitive and neural processes. Given that critical parts of the brain, such as the prefrontal cortex, are probably not fully functional until around age five (and continue to develop for much longer), there seems to be an obvious explanation at hand for childhood amnesia. But are delays in neural and cognitive development the only explanations we can look to? In trying to understand any boundary—in this case the childhood dividing line between memory and forgetfulness—it's a good idea to get as close as you can to the boundary. The territory of the past is lost to us as adults, but is it lost to us as kids?

The answer, of course, depends on the age group you are investigating. Some studies have shown that if you quiz children rather than adults about their earliest memories, the boundary of childhood amnesia seems to shift. In one recent study, researchers in New Zealand recruited four groups of participants: five-year-olds, older children (aged eight and nine), adolescents (aged twelve and thirteen) and young adults. For each participant, the researchers created what they call a Timeline: a horizontal line depicting different years of the individual's life, with photos of the participant attached at some of the ages.

Once the participants had shown that they understood what the Timeline signified, they were asked about their memories of first a recent and then some more distant events—specifically, an event from age three, one from before age three, and the earliest memory that could be brought to mind. Dating the event was a matter of indicating on the Timeline where the memory fit in; these reports were also confirmed where possible by parents. Participants were additionally asked about events from those times that their parents had nominated as being particularly significant.

The findings supported the idea of a shifting boundary of childhood amnesia. The children (both younger and older groups) reported a greater proportion of pre-three memories than did the adults. More than 20 percent of the memories of the children dated from before their first birthday. Most of the children's memories were pre-three; most of the adolescents' and young adults' were post-three. When the researchers looked at the very earliest memories reported, they found that the average age of the earliest memory was younger for the children: between one and two years, compared with about three and a half for the adults. The memories of children, adolescents and adults were equally episodic in quality, suggesting that children were genuinely recalling events and not simply parroting stories they had heard about their lives. The conclusion is obvious: children are not as amnesic about their early lives as adults are.

The youngest children in the New Zealand study were aged five, which is well into the period in which autobiographical memory is thought to be fully functional. What if you pick even younger participants, and ask children who are close to the age of offset of childhood amnesia? Ideally, we would have a way of examining memory that didn't rely on children's ability to put the memory into words, but no one has yet thought of a good way of doing that. Instead what we can do is ask children about their pasts as soon as possible after they have reached the level of language skill necessary for talking about them. Catching our research subjects young is also another way of getting around the problem of mediation: the fact that memories are as much rerememberings of things that we and others have remembered previously. The younger we get them, the more sure we can be that their memories are "pure" recollections of an event.

This was the approach taken in an intriguing recent case study. The author, a developmental psychologist and therapist named Aletha Solter, had been working with the family of a little Californian boy, Michael, who at the age of five months had had a short stay in the hospital while he underwent cranial surgery (to correct a deformation of his skull). The therapy he was receiving was aimed at relieving the behavioral symptoms of traumatization, including disrupted sleep and night terrors. Michael had had no further experience of hospitals since his stay as a baby. Solter took advantage of this fact in planning a study of Michael's memory for his time in the hospital. Crucially, she got Michael's family to agree not to discuss his hospital stay with him. In this way, she could ensure that anything he remembered could not be contaminated by previous recollections of the event, including details suggested by other people.

Solter then conducted two follow-up interviews in which she asked Michael about his memory of the event. The first interview, which took place when Michael was two and a half, showed him to have some strikingly detailed memories of his stay in the hospital two years before. He recalled that his eyes had been closed for a time (as a result of the surgery), that the nurse had been wearing a red blouse and scarf, and that his grandfather had sung the carol "Silent Night."

At the second interview, conducted eleven months later, when

Michael was three and a quarter, the picture was very different. This time, the little boy appeared to have no memory of his time as an in-patient. When Solter prompted him about the details he had recalled a year before, he had entirely forgotten them. As a two-and-a-half-year-old, Michael had limited, but detailed and accurate, memories of this event from his infancy. At three and a half, those memories had vanished.

These findings are intriguing for a number of reasons, but not least because they demonstrate that the two-and-a-half-year-old Michael was able to use language to describe events for which he would not have had the relevant language at the time. Although he was noted to be a verbally precocious child, at the age of five months he would presumably have had no language at all. Solter's study of Michael is not the first to have shown that children can later apply verbal labels to preverbal experiences. In one study, researchers showed two-year-old children an event (the activation of an interesting bubble-making machine) that was critically related to color (only a particular color of bubble soap activated the machine). Children who did not have color words at the time of the event were then, over a period of two months, given instruction in using color words. When they were tested for recall of the original event, a significant percentage of children who did not have color words at the time of the event nevertheless used their new color labels to recall what had happened. These children were able to use vocabulary they didn't have at the time to make sense of their earlier remembered experience.

The weight of the evidence, though, points to very limited verbal access to preverbal memories. A major force behind childhood amnesia is undoubtedly that children are trying to relate in language events that happened before they had language. Although Michael subsequently forgot his trauma, his case shows that there may be a sensitive period for access to such memories. If your language is good enough in that period between about two and three years of age, you may be able to cast a line back into babyhood and use your newfound words to capture events for which you had no words at the time. That may give you a chance to rehearse, consolidate and talk about the memory in years to come, in such a way that it follows you into adulthood. If your language is not as advanced as Michael's, you will lack that thread on which to pull. For these children (which is most of us), the joys—and terrors—of infancy are lost forever.

WHEN I FIRST ASKED ATHENA about her memories of babyhood, she would have had good reason to wonder what I was asking her. We were in Sydney, at the end of our six-month sabbatical there. She was nearly three, around the age at which childhood amnesia is supposed to diminish. But I was skeptical about the likelihood of her remembering much about her life to now. When she told me, quite firmly, that she could remember nothing of her infancy, I assumed that it was evidence of a memory system undergoing rapid change. Many factors could have led to forgetting at this time. Her ability to encode information would have been steadily improving, meaning that more recent events would have had a better chance of being laid down, but also that more distant events would have been less likely to leave an enduring trace. Her efficiency at consolidating and storing these memories would also have improved quite dramatically, although possibly too late to save memories from the more distant past. As she got older, and those critical memory structures in the medial temporal and frontal lobes began to reach maturity, the points of weakness in her remembering would have shifted from encoding and storage to retrieval. What limited her memory now would have had more to do with the ability to retrieve memory traces rather than with the ability to lay them down in the first place.

With the right cues, though, she proved herself to be at home in her past. When we returned to England, the three-year-old Athena was surrounded by distantly familiar things. Out of a choice of three distinctively patterned napkin rings, she was able to pick the one that had been hers. She remembered helping me to prepare breakfast in bed, an amusingly bohemian weekend ritual established before we left for Australia. Neither of these details had been rehearsed or talked about in the interim. The boundary of childhood amnesia was permeable, and when the conditions were right and the cues familiar, she could be inspired to recall.

Perhaps her problem was not so much with remembering per se, as with remembering under her own steam. If, out of the blue, you asked her about her past life (as I did that day in Sydney), she had an immediate challenge of navigating that knowledge, of orienting herself within it. But if she had cues from other people, or stimuli in the world around her that could trigger her memories, she could remember. She understood the question I was asking her, but she didn't know how to organize memory for herself. At three, she was only beginning to have a life story, and so it was no surprise that she wasn't yet very good at telling it.

She did actually remember something for me that day. After repeatedly informing me that her infancy was a blank to her, she eventually had an assertion to make about it. Crucially, it wasn't about the bald facts of the time, but about her own experience of it. "It were very sunny," she said. As a clue to the past, it wasn't much to go on. But it gave me an idea about what that past meant to her. Whatever image was occurring to her, she had remembered some of its sensory qualities.

These bright shards of memory tell us a great deal, I think, about how memory works, and in particular about how our memory reconstructions are built up from sensory and perceptual fragments. In fact, vivid visual details are among the most prominent of the sensory-perceptual experiences that form the raw material of memory-making. Neurophysiological studies show that they are stored in areas of the brain far removed from the frontal and medial temporal lobe regions that perform the task of assembling them into actual memories. Most memory researchers believe that the sensory and perceptual elements of memories are stored in the same parts of the brain—the different sensory cortices, such as the occipital and temporal lobes—by which they were first processed. The bits of a visual memory, for example, are stored in the visual cortex, and only assembled, when necessary, into a full memory in the hippocampus and related structures. Before autobiographical memory becomes fully organized (thanks in part to the maturation of these frontal and medial temporal lobe brain systems), these sensory-perceptual elements persist only as fragile fragments.

Many early memories have these qualities. In Vladimir Nabokov's autobiography, *Speak, Memory*, he sees "the awakening of consciousness as a series of spaced flashes, with the intervals between them gradually diminishing until bright blocks of perception are formed, affording memory a slippery hold." Recalling his epiphany of nascent self-awareness on the occasion of his mother's birthday, in late summer at his family's country estate in Russia, he describes his memory as being invaded by strong sunlight, "with lobed sun flecks through overlapping patterns of greenery." Consider, too, A. S. Byatt's early memory of herself as a small child looking over the wall into the playground of a primary school. She recalls the stone from which the wall was made as flaking into "gold slivers," the sun being very bright, the leaves of a tree catching the light and turning golden, the blue sky containing a huge sun.

Many of these light-filled mnemonic images are also judged by their owners to be their first memories. The painter Georgia O'Keeffe, famous for her luminous images of the New Mexico desert, claimed that her first memory was "of the brightness of light—light all around." In fact, scientific research suggests that such images may be more likely to qualify as first memories than those moments of dawning self-awareness recalled by Wharton, Byatt and others. The Nova Scotian researchers who asked participants about their earliest event memories also asked them to recall what they called "fragment memories": isolated shards of memory without autobiographical context and often taking the form of an image, a behavior or an emotion. The ages at which these impressions were judged to have occurred were significantly earlier than the event memories, leading the researchers to conclude that memory begins, at the offset of childhood amnesia, with fragments of experience rather than representations of complete events.

Sometimes the perceptual and the self-conscious coincide in early memories. Two of the most famous accounts of earliest memories come from Virginia Woolf, in her essay memoir "A Sketch of the Past." In the first, Woolf describes an image of "red and purple flowers on a black ground": the pattern of the anemones on her mother's dress, seen from close up on a train journey to St. Ives. The second memory relates to the St. Ives visit itself, and she convinces herself that it is actually her earliest memory. She remembers

> *lying half asleep, half awake, in bed in the nursery at St. Ives . . . hearing the waves breaking, one, two, one, two, and sending a splash of water over the beach; and then breaking, one, two, one, two, behind a yellow blind . . . hearing the blind draw its little acorn across the floor as the wind blew the blind out . . . lying and hearing this splash and seeing this light, and feeling, it is almost impossible that I should be here; of feeling the purest ecstasy I can conceive . . .*

Some writers claim absolute certainty for their early memories, but Woolf is refreshingly honest about her vagueness about certain details: the question of whether the journey unfolded on a train or an omnibus, whether she was traveling toward or away from St. Ives. She is also open about her uncertainty about which of these memories came first, relying instead on detective work about the evening light in the train carriage and the inference that it must therefore have been on her return to London, rather than on the morning journey out to St. Ives. Because they are not integrated with other kinds of autobiographical information, and because they do not carry date stamps of their own, memory fragments are hard to pin down in time. Vivid as it is, my image of myself as a toddler pushing a forklift truck along the carpet contains no information tag that can date the event. If I want to try to place the memory in time, I have to use inference and detective work, just as Woolf did, such as trying to identify in which interior the scene of play occurred.

It is perhaps no accident that our earliest memories are full of light. Sensory-perceptual information is the raw material of episodic memory, and visual sensory information is the most salient of all. For the writer and naturalist W. H. Hudson, such fragments were "isolated spots or patches, brightly illumined and vividly seen, in the midst of a wide shrouded mental landscape." As the processes of memory become more integrated and organized, these sunny fragments gradually cohere into autonoetic, narratively structured retellings of past events. But our memories remain dependent on these bright raw materials, such as can sometimes take us back into the earliest days of childhood.

5

WALKING AT GOLDHANGER

MEMORIES OF CHILDHOOD are often memories of terrors. For me, the tidal mud in the estuary of the river Blackwater was as menacing as any fairy tale monster. From the boatyard jetty, as wobbly as a rope bridge strung between the mud-locked boats, it was a green-tinged taupe color, fringed with clumps of weed. Its slimy convexities were pitted with worm holes and veined with coarse, haphazard channels where the tide flowed thinly back to the sea. It stank of the sea, of brine and marine life, but it also had its own oily musk, suggesting life-forms that had died, rotted and then been buried, more years ago than I had been alive. Out on the mudflats, at low tide, it gleamed as brightly as the water beyond. Dad used to tell me that a child falling into that mud would never be seen again. Even as his words went through me, I had my doubts. The mud looked too solid, too well set. Dad's hairy, red-tinged bulk was another matter, but the mud would surely resist the impact of a lightweight like me. Tempting as it was, I never got the chance to put his words to the test. The mud was everywhere, but it was also fatally out of reach. When we launched the dinghy to take us out to the mooring, we stepped from the wet concrete of the runway straight onto the yielding rubber of the inflatable. I would have had to shimmy down one of those rust-dark jetty posts to dip my toe in it or, more improbably, contrive to fall in. Its unreachability made my fascination all the more extreme. It could affect me, but I could not affect it.

After his divorce, my dad had found a welcome distraction in sailing. A friend of his, a whiskery schoolmaster named Monty Stanley, owned a twenty-foot Bermuda sloop called *Doralind*, which he wintered in Maldon and moored for the summer at West Mersea, a little farther along the Blackwater coast. Dad had no investment in *Doralind*, but in return for his work on the boat he got to sail with Monty whenever he wanted, and to bring us kids along when we were with him for the weekend. I was exposed to the mud early on. I would see it in the boatyard at Maldon, where on winter weekends we would be painting the hull or trying to pry open tins of varnish with a screwdriver. The tide always seemed to be out, as though in conspiracy with my obsession. I wanted to know how deep a boy of my weight would go. Ankle-deep? Knee-deep? Would I really sink from sight completely? How many children, in Maldon's long history, had been muddily snuffed out in this way?

It trots off the tongue, what Dad said to me about the mud, and the complex fascination it inspired. It is the stuff of memoir: a vivid childhood memory freighted with significance and now seeming to contain an incontrovertible truth about one's own life. But I have learned too much about the fallibility of memory to be taken in so completely. I'm pretty sure about the sensory impressions: the stink of the river bottom; its gleam under a reflecting sky. I don't doubt the remembered cries of black-headed gulls, the constant *tink-tink* of lanyards against masts. I'm confident about the part of the memory that relates to facts about my life, those autobiographical details about what I was doing at what point in time. It was after the divorce and before I started at boarding school, which puts my age as between six and ten. But that poignant story-opener,

"My dad used to say . . ."—how much of that is of my own making? I have no memory of him actually saying those words. In no version of the mental screenplay is he mouthing them. What I remember is the thought: *You will disappear forever.* Maybe what I want to say is that I *know* that my father said it, like I know that he was born in Newcastle-under-Lyme and that his last car was a Nissan Primera. But I might just as easily have put that warning into his mouth. It was my terror, after all, my fear driving the remembering. He is there in the memory, but his feelings are not propelling it. Memory serves its own master. It doesn't work for anyone's purposes except the rememberer's. I am going back to this moment on my own terms, not anyone else's.

This is how easy it is to be drawn into the fiction. If we are to be honest about how memory works, we need to be wary of its charms. In my memory of our sailing days, I suspect that I am recalling the thrilling terror of the mud, along with the emotional quality of being with my dad. In the laboratory of memory, that conjunction is converted into an image of the man saying the words. My dad might never have said what I have attributed to him, but I have created a fiction in which he did. It makes for a good story, and a vivid mental image, but it is not necessarily—in any objective, timeless sense—"true."

The problem I face in recalling my father's words is a familiar one for memory researchers. Memory can be surprisingly inaccurate when it comes to the little details, such as the exact words that a speaker used. That's because it tends to exert its grip on the meaning of what happens to us, and lets surface details slip. Remembering anything depends on the process of *encoding*, the conversion of the relevant information into a code that is recognizable to the memory system. Other processes, such as storage and retrieval, are critical as well, but without encoding, memory would be a nonstarter. Everything you remember, from your first day at school to the name of your last lover, has at some point been encoded into a form that your brain can use.

Thinking about encoding helps us to understand why remembering people's specific words is so difficult, and it also takes us right to the heart of how memory researchers go about their business. Generally speaking, the more deeply information is processed at the encoding stage, the better it is remembered. This finding, replicated in countless psychological experiments, has become known as the *levels of processing* effect. In one classic study, participants studied lists of words under different conditions, without being told that they would later be asked to recall them. One group of subjects had to answer questions about the physical properties of the words presented, such as whether they were printed in capital letters. Another group had to make judgments about how the words sounded, such as whether they rhymed with the word *dog*. A third group had to answer questions about each word's meaning, such as whether it would fit into a particular sentence. In each condition, the amount of processing, or meaning extraction, that the subjects had to do increased by one more step. When it came to an unexpected test of recall for the lists of words, significantly more was remembered by the participants who had had to process the information more deeply. Although many other factors influence whether information is remembered, the levels of processing effect is a telling clue that memory is meaning based.

So it is with more complex verbal structures as well. When we recall what a person said, we recall the meaning of their words, the gist of what they said, rather than the verbatim information. That's because, when we listen to verbally presented information, we pay attention to (and therefore encode) the story being told rather than the specific word choices and syntactic niceties of the speaker. A sensible strategy, you might think. Perfectly formed their sentences might be, but most speakers are more concerned that you get their message than that you admire their prose style.

To date, we have understood little about how the brain cuts to the essentials like this. A first step in this direction was recently taken by some researchers at the University of Western Ontario, who designed an experiment that involved reading sentences to participants in three phases. In the first phase, the volunteers simply listened to the sentences, each of which described a unique, vivid scene, such as a comedian receiving a standing ovation after a performance. In the second phase, the participants listened to another set of sentences while making judgments about how pleasant the sentences were, or how well formed. Some of these sentences were the same as those that had been heard in the first phase. Others differed from those original items in having the syntax or grammar changed slightly (for example, a clause might have been moved around within the sentence, without affecting its overall meaning). A third group involved a change in meaning ("a comedian" being changed to "an actor"). In the final testing phase, subjects were given a recognition test for the new sentences presented in the previous phase. As the researchers had predicted, sentences that involved a semantic, or meaning, change were recognized much more accurately than those that were syntactically different. The participants had kept track of how the gist of the sentences had changed, but their memories were almost blind to the actual form of the words.

The explanation for this "verbatim effect" seems to do with how the brain processes and retains novel information. It is well known that we are more likely to commit new information to memory than information we have encountered before. You don't remember your seventh day at school or your ninety-second; you remember your first. But memory is not impressed by any old kind of novelty. It is specifically the kind of newness that changes the meaning of information, rather than its surface form, that has the best chance of being remembered. If this were true, then you would expect that new semantic information would specifically activate those parts of the brain known to be involved in episodic memory.

The Canadian researchers were able to address this question because they were scanning their participants' brains at the same time. They were looking for activity in that part of the brain, the medial temporal lobe, known to be deeply involved in episodic memory. We have already seen how important the hippocampus and its neighboring cortical structures are in creating mental maps of our physical environment. But that is only part of the story of this mysterious hornlike structure. The hippocampus also plays a crucial role in remembering the associations between bits of information and binding them together into an episodic memory. At retrieval, the return of one of those elements into consciousness can be sufficient to allow the hippocampus to complete the pattern of associations, allowing the other features of the memory to be retrieved. The hippocampus seems not to store the information itself—the building blocks of our episodic memories are distributed in various other locations throughout the cortex—but it does seem to store the associations between those elements of remembering. And it seems particularly interested in *new* associations, confirmed by brain-scanning findings that the hippocampus becomes particularly active when it encounters new material.

The Canadian neuroimaging results showed that a portion of the hippocampus on the left side of the brain responded specifically to semantically new, but not syntactically new, sentences. Information that changed the relationships between ideas—which changed the sentences' meaning or semantics—seemed to activate this part of the brain in particular. Not only that, but this information was also remembered by the participants more accurately, which is what you would expect given the proven importance of the hippocampus for memory-making. In order to be remembered, novel information has to trigger the cascade of activity in the medial temporal lobe that is involved in laying down a memory. The fact that it is only new semantic information that gets this part of the brain working accounts for memory's reliance on meaning rather than surface form.

It seems that there are some good scientific reasons for distrusting the memoirist's fondness for quoting remembered speech. Perhaps Dad said something like "that mud is very dangerous, like quicksand," and my imagination supplied the rest. Now that I think about it, I'm pretty sure I heard him telling horror stories about quicksand in contexts quite different from the sailing one. As a child, I made a strictly literal identification of quicksand with ordinary sand, such as you would find on a beach or in the desert. I had images of people and camels sinking slowly into the valley between two dunes, the shimmering air thick with their complaints as the greedy earth inched up around their legs. I suspected that Dad's own testimony was based on hearsay—he had not done much desert travel of his own—but that didn't stop my imagination from firing. Somewhere along the line I linked the quicksand nightmare with the warnings about the tidal mud, and the two fictions melded into a new one of memorable power.

Of course, I cannot be sure. If I were able to ask him, Dad would be no more accurate than I am at recalling the exact words that he used. He is no longer here, and that fact makes the detail more precious and its lack more poignant. Memory serves its master: I *want* to hear his words, and so they are obediently supplied. Remembering the words of the dead, like so much of memory, is at best a kind of informed storytelling. Maybe that's why, in trying to recapture those lost times with my dad, I can be most certain about a memory in which nothing was said at all.

IT IS AUGUST. LIZZIE, THE kids and I have just come back from a summer holiday in the heat of Portugal, and I have that disoriented, lost-in-time feeling you get after doing not very much for too long. It is a gray, mild day on the Essex coast, with a strong southwesterly rattling the lanyards of the boats docked in the yacht club yards. I have been used to blue skies and searing temperatures, and this ambiguous weather adds to the feeling that I have stepped back into a world that has changed subtly in my absence—in which, because I no longer have any useful function, I may as well continue to pursue indolent dreams.

The path out of Heybridge Basin is a metaled track running alongside a concrete seawall. The boatyards are tangles of disorder, clumps of brightly colored sheds jutting out from the seawall on brown stilts. The path takes you right through the hearts of the yards, quickening curiosity with a faint sense of intrusion. The tide is out, and dinghies lie deposited on the mud at uncomfortable angles. Beyond the boatyards, on the landward side, you come across a city of static caravans, three holiday parks merging into one sprawling, white-roofed conurbation. The flood barrier is made of concrete blocks with KEEP OFF WALL stenciled in red paint. Beyond it, the mudflats stretch out to a steely river, gleaming in the sunlight that falls between fast-moving clouds.

We would never have come this way when I was a child. Dad would have found it too crowded, with too many signs of human occupation. The caravan parks, for all their promise of a cheap holiday, would have filled him with dread. He might have admired the ingenuity of some of the inhabitants: fashioning a clothesline from an old boathook, a birdhouse from a beer crate, expressing their commitment to recycle, reuse and mend. Mostly, though, the prospect of all that housing meant the prospect of people. For Dad, a good walk was a solitary one. When he spent Christmas Day out here—as he usually did when we kids were with our mother, and he had no family around to bully him into the yuletide spirit—he would announce afterward that he had had the best Christmas Day ever. The solitude was part of the recipe. Walk for hours, have a sandwich and a flask of coffee instead of a turkey dinner, and see not a soul.

The landscape, then, explains its own unfamiliarity. I know, because my father's remembered foibles tell me so, that we wouldn't have started our walk from here. But I have a more powerful reason for excluding this landscape from memory. I don't recognize it. It's not just that I know—cognitively, in the way that you know a fact—that we wouldn't have come this way. It's that I don't remember coming this way. The scene is gleamingly unrecognizable. I don't know if I am actually responding in some positive way to the scene's newness, or simply failing to feel an expected familiarity—but I am responding to it anyway. I know I have not been here before because I *feel* as though I have not been here before. The feeling seems to guarantee the knowing, and it's not the sort of sensation you can argue with.

There are other reasons to think that this was not the way I used to come as a child. I remember—that is, I have a clear, vivid recollection—that we used to park in the village of Goldhanger and cut across fields to get to the seawall. The walk out of Heybridge Basin was never part of it. This is not a judgment based on the absence of familiarity, or even a positive feeling of unfamiliarity; it is a conscious recollection, a genuine, there-in-the-moment episodic memory. I have an image of Dad and me parking, leaving the car and setting off. I don't remember any details of the walk across the fields, but I remember the parking. I remember it because Dad was always anxious about leaving his car, a white Vauxhall Carlton station wagon, parked on a village street. I seem to remember an overgrown green, perhaps with a pond and some benches. There were no parking restrictions, no yellow lines that we could see, but still he would worry about leaving the car anywhere that wasn't his own driveway. I suspect that it's this anxiety that sticks in my memory. We remember the unexpected, and the sight of an adult showing any barely perceptible emotion was enough to burn the memory for posterity. This sense of my father's vulnerability was not exactly unusual; my parents had been through a divorce, and they were already painfully human to me. But still I must have been intrigued by this faint show of weakness, as he engaged the newfangled central locking system and walked on, checking back at the car one last time.

Two kinds of memory, then. I remember that; I don't *recognize* this. To find something familiar, you don't have to be able to call it to mind at will. You simply have to know it when you see it. In the human brain, the familiarity system involves a network of neural centers that includes the medial temporal lobes and surrounding structures. Throughout the animal kingdom, versions of this system turn out to be neurological circuits of exceptional power. In one experiment, pigeons learned to discriminate between pairs of images: random squiggles or photographs of natural scenes. When they were tested again a full two years later, the birds quickly relearned the associations. It was as though they could still distinguish the images they had seen two years ago from new, unfamiliar ones. If the same powerful (and presumably evolutionarily advantageous) force is at work in my own brain, I should be able to recognize much more than I can explicitly recollect.

I have brought no map with me today. This is partly because I want to see how much I remember, and partly because I figure that even I, with my feeble sense of direction, can't get too lost in this landscape. As long as I keep the river to my right and stick to the path along the seawall, the proper functioning of my hippocampal mental maps will ensure that I join up with our old route eventually. I know that Goldhanger is somewhere up ahead of me, because I planned my route on Google Maps before I left. Somewhere in that smoke-gray coastline is West Mersea, where we used to sail. All I have to do is be vigilant for the moment when the scene becomes familiar, when that primordial familiarity mechanism clicks into action and tells me that I am somewhere I have been before. The seam between an unfamiliar landscape and a familiar one will mark the boundary of my childhood walk. I can't imagine any stronger test of that feeling of familiarity. As soon as it comes in, I'll know I'm there.

At least, that's my plan. If I had managed to find my way to our parking spot in Goldhanger and followed our original route, I would have started with the expectation that everything I saw should be familiar. This way around, I have to rely on the feeling. But as the caravans drop away and the scene becomes wilder, the prospect of being able to rely on that feeling seems more and more unlikely. The concrete blocks have gone, and the seawall has become a rampart topped with a gravel track and clumps of yellow and blue wildflowers. Blackberries are ripening on brambles. There is a track heading inland, with a few brown holiday cabins threaded along it, but I don't recognize it as our path out of Goldhanger. It's a weekday, and no one else is out walking. Across the water I can see the thin, dark, wooded strip of Osea Island. It won't do as a landmark; it's visible from anywhere along this stretch of coast, and so not nearly specific enough. Closer in, there is precious little more to get my bearings by. On the landward side there is a water-filled canal like the moat around a castle, with thick green rushes growing up from it. Beyond that, cows have their heads down at pasture.

It certainly looks much like I expected it to. But the feeling of familiarity is not exactly assailing me. When someone with temporal lobe epilepsy is about to have a fit, they often have a sense of déjà vu: an impression that the moment is familiar, without them quite knowing why. In epileptic déjà vu cases, this feeling may be a forewind of the electrical maelstrom that is to follow in the temporal lobes. The neuroscientific evidence suggests that the feeling of familiarity relies on activity in a network of regions centered in the perirhinal and parahippocampal cortices (those bits of cortex that closely abut the hippocampus). To be familiar with a stimulus, you only need to *know* that you have seen it before; you don't have to have an episodic memory, or recollection, of having encountered it on a previous occasion. That separate process of recollection (or what we ordinarily call *remembering*) is underpinned by a separate neural system, centered in the hippocampus and involving several other regions of the medial temporal lobe system including the fornix. These two neurally distinct processes provide two routes through which an environment can be judged as having been encountered before. You can recognize a place because it is familiar (because activity in your perirhinal and parahippocampal cortices tells you so), or because you actively recollect an occasion on which you were previously there (in which case it is your hippocampus and associated structures that are busy).

I have no idea whether the pigeons that studied all those photographs had the same rich sensation of familiarity. All I know is that that feeling, for me, is today in short supply. What's out here makes a beautiful, bleak scene, but it doesn't feel *mine* in any way. Whatever a memory is, you surely always know to whom it belongs. As the wind strengthens, scouring my face and hurling the scene at me with greater force, I begin to wonder whether I've got this hopelessly, spectacularly wrong. A routine of my childhood, a precious thread of my past with my dad, has left little trace. I've always remembered it as memorable, but it is not. Memory can trick you and let you down in all sorts of ways, but simple forgetting is the plainest disappointment of all.

Perhaps it has just been too long. I know for certain that I haven't been here since Dad died, eleven years ago. I have been living at the other end of the country, and trips to Essex have been exclusively for family visits. I spent most of my youth trying to escape my home county, and this is the first time I have shown any interest in it. From the mid-1980s onward, Dad was spending half of each year in India. Before that, in 1980, his life changed when he met his new partner, Ann, and our weekends together were mostly spent with her and her family. The walks I'm remembering happened when I was a child at boarding school, between the ages of ten and twelve. Dad would pick me up after chapel on Sunday, and drop me back at my boarding house in the early evening. There wasn't time to go sailing or work on the boat, so we would go for a walk instead. I'm relying on information about our lives as much as on my own ability to remember, and it makes me pretty certain that I have not walked this seawall for twenty-eight years.

That's a long time for forgetting to get to work. I'm not saying, of course, that I remember all those days when I *didn't* come here with Dad. You can't have a memory for not doing something; you can only *not* have a memory for doing it. I cast out my net and it comes back empty. However it is that recollection works, it gives us a basis for deciding whether something happened or not. In science, we say that the absence of evidence is not evidence of absence. Memory works differently. If we can't retrieve any memories of a thing happening, our best guess has to be that it didn't happen.

In working out whether something happened or not, we don't always have to go through a long process of casting back through our memories, or what psychologists call doing a *serial search*. Have you ever had dinner with a pop star? I suspect that you can answer that question straightaway, without much thought at all. You don't have to trawl back through all the people you *have* had dinner with, looking for a pop star and finding none. If you have dined with a pop star, the incident would probably be important enough to be transformed from a single episodic memory to a fact about yourself (*You know me, I'm that person who once shared sushi with Lily Allen*). Episodic knowledge would have become semantic knowledge, and you would have been able to answer the question as quickly and as certainly as you'd have been able to answer a question about your height or your shoe size.

So there is a lot of detective work, both semantic and episodic, involved in answering questions about your own past. In remembering our walks at Goldhanger, I draw on biographical knowledge about the people involved as well as my own specific episodic memories. That is, I combine a kind of autobiographical *semantic* memory (which tells me the facts of my past life) with autobiographical *episodic* memory (which allows me to relive certain moments of experience). I use logic, and I make inferences. I use certain facts about the person I am to answer questions about who I have been. There are other places I have never been to, such as Stockholm and Addis Ababa, and I know those facts without having to trawl through a list of all the places I *have* visited. The timing of my last visit to Goldhanger is not as certain, but, in trying to recall it, I'm using knowledge about myself as much as I am relying on specific memories.

Today, in fact, that kind of biographical detective work turns out to be more productive than trying to spot the transition to familiarity. As I leave the track with the holiday cabins behind me, and see that there are no other paths that could have brought us here from the village, I realize that I must have been wrong to reject it like I did. I didn't know for sure that it *wasn't* the path we came along, and now, glancing back at it, I'm wondering whether it doesn't look just a little bit familiar. Wasn't that the same bepuddled, tree-lined track where, clutching my grandfather's old

First World War binoculars, I used to dream about the birds I would spot that day? In trying to connect that remembered reality to this new experienced one, I will clutch at anything. The motivation to tell a coherent story is a powerful one, and it pulls together facts that shouldn't necessarily go together.

The truth, though, is that the memories are not flooding back. I know that I am on the path I walked along with Dad nearly thirty years ago, but I know that fact mostly because I have worked it out. The past is too distant a country for me to get there under my own steam. But once logic has put me there, the feeling of familiarity is not far behind. Semantic memory makes episodic memory possible. The *idea* that this is the walk I used to take can be as effective a cue to recall as any specific familiar detail. Or perhaps I should say that it makes half-familiar things more familiar, and heightens the power of the surrounding cues to trigger actual episodic memories. It pushes up the faders of familiarity, and makes remembering more likely. In terms of the brain circuits that underlie them, familiarity and recollection may rely on different systems, but in practice one must feed into the other.

The wind out here is phenomenal. It howls into my face, makes my contact lenses judder. I'm spattered by a constant blast of sea fret. My T-shirt is blown back behind me in a shark's fin of flapping cotton. I spot a sloping bit of seawall, built from pale yellowish concrete blocks, and I duck down onto it. Immediately the battering stops. It's as though there were another world folded inside this one, with a drastically different climate, and I've just stepped into it. I shrug off my rucksack and sit down on the concrete incline. I look out at the water and see a brown-sailed barge gliding back to Heybridge Basin. The sudden calm, and the feel of the rough concrete under my hands, suffuse me with a faint, indescribable warmth. My perceptions seem suddenly sharper. It's true: there's a moment trapped inside this moment, and it is familiar to me. It makes me realize that this, or some other bit of seawall like it, is where we used to stop to have our picnic. We would sit out here in the wind, coats folded beneath us, and eat our whole wheat bread sandwiches and drink instant coffee from plastic mugs. These are still details of semantic memory, of autobiographical knowledge; I know the menus of our picnics as a fact, rather than as a memory. But this peaceful moment is bringing knowledge to mind that could take me somewhere. Connections are being made.

Which, after all, is what I've been hoping for. I've been expecting that the known will trigger the familiar, and the familiar will trigger the recalled. Familiarity and recollection are the two foundation stones for knowing where we have been. If I know anything about how remembering works, I know that the details of this windblown scene are the ones that will unlock the memory for me. I've zeroed in on a particular location, and I've done that because I want to see what surrounds me there. The answer is out there on the water, in the blurry smear of Osea Island, not inside my own head. Crucially, I know that the details that confront me now will be the same as when I was here last. In memory, context is everything.

There's a simple reason why this fact is so important, and it goes back to the principle of encoding. Whenever information is encoded, associations are made between the thing to be remembered and the cues that are around at the time. That's why context is such a powerful cue to memory. It is a well-established finding that we are better at remembering events and information when we are asked to recall them in the same context in which we laid the memories down. In one experiment, deep-sea divers were asked to recall lists of words they had learned underwater. They did better at recalling them in the same context as the one in which they had learned them (several meters underwater) than they did on dry land. In criminal investigations, it is standard practice to take witnesses back to a crime scene to help them recall. Remembering is in large part a fortuitous matching of the ambience of encoding with that of retrieval. Psychologists have argued that this is because the cues that are around at the moment of encoding (the laying down of the memory trace) are stored along with that remembered material. Consequently, the reappearance of those cues can make the memory bloom into consciousness again.

That's why I've come to this specific place to remember my dad, and not to any other. If I re-encounter the cues I shared with him (and only him), he will start to take shape in my memory. If it's going to happen, it will happen now, in the scene before my wind-blurred eyes. The light is on the water. Osea Island is a mysterious, uninhabited realm in the distance. Some diving birds are racing in formation across a silvery inlet. There is the oily smell of the greenish mud. Everything is in its place . . .

I feel the memory before I know it in any ordinary way. The drifting, lost-in-time feeling with which I set off from Heybridge Basin has become a sort of vulnerability, an acknowledgment of human smallness. I feel as a child would feel, snatched for a few brief hours from a controlling, organized regime, to be confronted by this bleak, massive world. I've been thinking hard, trying to fit these two realities together, but actually I should have been listening to my heart. In certain circumstances, an emotional state can trigger an episodic memory as powerfully as a landscape can. If my feeling at the moment of retrieval matches up with how I felt at encoding, the memory can be unlocked. If the cues are aligned—even if they are purely emotional cues—the recollections will flow. My feeling a particular way will be my escort back to the time.

Actually, I can't be sure what triggers the memory. I think it's the feeling, but the feeling is in turn triggered by the sensory context: the light on the water, the diving birds, the smell of the mud. All I know is that I am eleven years old again, or ten, or twelve. The same brown sailing barge is drifting across my sight. I'm imagining what it would be like to borrow a boat and row across to Osea Island. I have been reading *Swallows and Amazons*, and the creeks and channels of the Blackwater have been mapped out in my imagination—and, when I get hold of some colored felt-tip pens, sketched out on some sheets of scrap paper from the bottom drawer, with Dad's unwanted business letters Xeroxed on the back. I have waterproofed my hand-drawn charts, ready for adventure, in a plastic document wallet I filched from the stationery cupboard. This morning we stood in the kitchen and made our sandwiches: a smear of Flora, another smear of Marmite, some thin slices of cheddar, some fragile, membranous loops of Spanish onion. Decades later, when he is dying, I make him the same sandwich, but this time cutting off the crusts. There is a memory within the memory, and it stretches into the future to join up with another, then loops back to where it began. Dad is next to me, unwrapping the picnic. I feel his presence, the funny jolt of being with him, without sensing him physically. He doesn't speak. We spend most of these walks in silence, listening to the wind, the birds and the sound of our own breathing. It's as though he wanted to teach me about the spaces between conversations, not the conversations themselves. To be silent, unfathomably closed on matters of the heart, as he himself always is.

It's not that I don't hear his voice in other memories. I remember the gist of things he said, if not the verbatim details. Things that had an emotional charge; utterances that struck an embarrassed chord, that showed up his vulnerability or mine. The two of us at Danbury Lakes, Dad telling me, clearly anxious, about a big business deal he was going to have to try to finalize the next day. In the car park at my boarding school, just before the Sunday-night drop-off, quizzing me gently about the bullies. Walking by the canal one day, awkwardly but honorably offering to answer any questions I might have about sex. At Goldhanger, though, I can only remember silence. We have five hours to go before I have to be back at school. Surely this is the time to say everything that needs to be said, to broach the forbidden topic, unearth the truth that will explain everything? Whatever it is, we don't say it. We are comfortable with silence. Without the pressure of language, actual or expected, we can stretch out a little more easily into the people we are.

The binoculars on their ancient leather strap are heavy around my neck. Dad has a modern, lightweight pair with a plasticky, new-camera smell. Their big lenses flare into purplish disks when they catch the light. They have two pairs of rubber lens caps that I have to be careful to store safely in the binoculars' pigskin case. There is the taste of coffee in my mouth. I don't know when I started drinking coffee, but I did. The light on the water is beautiful and fragile. Something unnameable is tugging at my heart. I watch the birds darting over the water and try to remember what they are called. Greenshank. Redshank. Something Wagtail. A Something Something Grebe. I should know their names, every one of them; they are printed out in my bird-spotter's inventory, and I have ticked them all off. But there is no bird-spotter's inventory. There are no binoculars. I have forgotten everything, even the facts. I can taste the coffee, but its flavor is faint and unremarkable, and I have been drinking it all my life.

6

NEGOTIATING THE PAST

WHEN MY DAUGHTER, Athena, was seven, we went back to Sydney and revisited some of the places that had become familiar to her during our sabbatical there in her third year. For the most part, the cues that I thought would be most redolent for her, such as the sight of the Harbor Bridge and the Opera House, did little to connect her to the past. Instead, she recalled some odd, trivial details: a moment on a trip to the Blue Mountains when a connecting door to her hotel room wouldn't open because a radiator was blocking it; an afternoon on the beach when she made a spectacular mess of a chocolate ice cream. I remember wondering at the time what they told me about her, these weird bits of flotsam that had reached the surface of her forgetfulness. They were charming and unpredictable, and they spoke of the distance between us. They also showed me that she was making sense of her experience differently, that things that mattered to me did not matter to her.

Coleridge called them "worthless straws," and pitied the fact that such fragments alone "should float, while treasures, compared with which all the mines of Golconda and Mexico were but straws, should be absorbed by some unknown gulf into some unknown abyss." Fragment memories are not worthless, of course; they are cherished by most of us, although it is hard to know what they mean. A Freudian might tell you that these innocuous images are "screen memories," put up to shield painful truths about our early lives. My forklift truck memory, for example, might be screening the memory of some repressed plot point in my early psycho-sexual drama. Most modern memory researchers would discount this interpretation. They would argue that early memories are fragmentary simply because they have not yet had a chance to become organized into structured systems of remembering. There are many reasons for this, as we have seen; many systems that need to be in place before autobiograph-ical memory can get going. But even Coleridge's "worthless straws" can give us some clues to the kinds of factors that might be important in this process.

One thing we can say about the seven-year-old Athena's memories of Sydney is that they would have been talked about. These moments had been allowed to become organized because they had been the topic of conversations with others who were present at the events. The radiator incident happened during a visit to Australia by Athena's grandparents, who took her to the Blue Mountains while I was off on another trip. The ice-cream event was part of a family outing to Dee Why, one of Sydney's northern beaches. There would have been lots of chat about these amus-ing happenings (as there also would have been about more traumatic events), and therefore lots of opportunities for beginning to integrate them into memory.

It turns out that children are willing to talk about the past from quite an early age. As a toddler, for example, Athena was able to talk about events that had happened quite far back in her past. At eighteen months, in a phone conversation with her mother, she could talk accurately about a visit from her godfather, Rhett, a few weeks earlier. She specifically seemed to remember how she had said good-bye to him as he left on a train: "Rhett-train-bye-bye!" Two months after that, she was still talking about the fluffy duck rucksack he had brought with him. Asked what she wanted to read one bedtime, she said, "*Peepo!*," referring to a favorite board book. Then she told us, "Book hurt cheek," recounting an incident from two months earlier when she had tripped and fallen against this same book, hurting her face.

Such anecdotal observations are backed up by more systematic studies. Researchers have shown that, for most children, the first references to past events typically occur from about eighteen months onward. Two-year-olds can give simple responses to parental inquiries about past events. With the right support, they can cast quite far back—six months or more—into their own histories. This leaves us with the paradox of children being willing to talk about the past before they can properly remember it. Amnesic they may appear to be, but they are not shy of talking about days gone by.

That seems less of a paradox when we look more closely at how autobiographical memory works. Toddlers are not amnesic; their problem, if it can be described in that way, is that they are insufficiently skilled at organizing their autobiographical knowledge. There is now strong evidence that the development of this organization is partly a social process. Some of the most telling findings come from studies of parent-child conversations about the past. Parents differ in their approach to these conversations, with some taking the opportunity to elaborate on the topic of conversation, by providing orienting information, such as details on location and characters involved, and evaluative information, emphasizing the emotional significance of what happened. Other, less elaborative parents keep their contributions at a more factual level. In turn, children who are exposed to this kind of elaborative input produce richer memory narratives of their own, when tested later on in childhood, than those who are not. A recent study suggests that the effects of parental elaboration can be long-lived. Adolescents who have been exposed to an elaborative conversational style in the preschool years produce earlier memories than those whose parents have tended just to repeat factual details. A certain kind of parental talk seems to make the past stick better in children's minds.

Lizzie and I used to try these ideas out in our own kids' bedtime routines. Before the age of about five, when they typically begin to be able to produce their own autobiographical narratives, it is not unusual for Western children to engage in quite frequent adult-supported conversations about the past. We played a game called "What We Did Today," which involved us discussing the events of the day and trying to encourage the children to provide elaborative detail. It was a social process in which the children's remembering was heavily scaffolded by our own contributions. In a way, it was just another version of storytime, but in this case we were constructing a story for ourselves, about ourselves. It was part of our process of creating a family mythology, and ensuring that we would remember the days as they hurtled by. Even now, the kids still find it a comforting if occasional ritual.

To remember the past, you tell a story about it. And in recalling the memory, you tell the story again. It's not always the same story, as the person telling it does not always want the same things. Memory fits in with the demands of the present as much as it tries to remain faithful to the facts of what happened. It incorporates new ideas, including snippets of information that have nothing to do with the original events. As children become better storytellers, they become better rememberers. But their memory system also becomes more susceptible to distortion, as it sucks up other facts and convinces itself that they were part of the memory. As Virginia Woolf noted, there is something particularly pure about early fragment memories, detached as they are from bodies of information that eventually weaken their power. "Later," she wrote, "we add to feelings much that makes them more complex, and therefore less strong; or if not less strong, less isolated, less complete."

As memory becomes more organized, it becomes more reliable. The more firmly a memory trace is laid down, the more distinguishable it becomes from other memories, and the more likely it is to be integrated with information that is relevant to the self. The features that make it distinctively and persistently part of one's *own* experience are more conspicuous. High-quality memories like these, well integrated with other sources of information, are easier to recall. But, paradoxically, they also have a greater potential to become disconnected from what actually happened. As memory leaves the world of fragments and unintegrated emotions, it also becomes more prone to distortions. The more memory becomes organized, the more slippery it becomes.

"WHAT DID GRANDAD PHILIP ALWAYS say?"

Isaac is doing some online shopping. He has some leftover holiday money to put toward a new game for the Wii, and I am trying to help him to work out whether, with a couple of weeks of pocket money thrown in, it will be enough.

"Do I *want* it?" His big sister starts enumerating criteria on a thumb and two fingers. "Do I *need* it? Can I *afford* it?"

I'm not sure that my dad always satisfied these three conditions when he went shopping. But his rule for parting with cash has become part of the children's own decision-making processes. It is one of the mantras they know him by. He died more than a decade ago, too soon to get to know any of his grandchildren: Athena and Isaac, and their cousins Lucy and Annabel. As time has gone on, I have wondered more and more about how the children are to know him, how I myself should talk about him, and the rights and wrongs of negotiating the memory of someone who is no longer here.

Our memory of him is not a particularly visual one. We don't spend a lot of time, as a family, going through photographs, and Dad died before digital pictures and video became ubiquitous. Talking to the children about Grandad Philip's funny pronouncements means that he becomes more real for them than a photographic image. It allows the kids to own a bit of him, to incorporate him into their way of looking at the world. The stories of his outraged behavior in restaurants and hotels, his subtle ploys for getting a drink when he needed one, have become real for the children, too. I want to say that they remember this affectionate, vulnerable, opinionated man, even though their stays on the planet did not overlap.

It's a harmless idea, surely. Grief and regret for what has gone, and pride and joy at what has arrived to take its place: all these emotions have combined to help me try to fix that broken link between the generations. I can't be the only parent who has tried to implant a child with a memory for a lost grandparent or, more tragically, for a dead parent or sibling. But something makes me uneasy. I am actively manipulating their take on the past, tampering with what I should leave alone. Among all the murky choices parents have to make, this one is rarely examined. Even if it were possible to seed a memory in this way, what kind of memory should it be? An honest, warts-and-all depiction, in which the dead person is anything but perfect? Other reveries of lost family members can be too good to be true: those grandmothers of hazy memory too often did nothing but clutch children to their downy chins and bake exquisite cakes. In attachment research, romanticizing past relationships in this way is taken as a sign that important failings in those relationships are being swept under the carpet. When we are remembering the dead, honesty seems the healthiest policy.

I also need to ask what kind of entity I am asking the kids to remember. They sometimes hear me saying things like, "Just because you can't see him with your eyes, it doesn't mean he doesn't love you." It's as though the detail of him being dead were no obstacle to their getting to know him. I'm asking them to have a relationship with someone who can't be seen or heard, who never takes his turn for child care during the school holidays or calls to find out how the cricket match went. At least Santa Claus brings them presents once a year. The difference is that there is some documentary evidence for Grandad Philip's existence. The children can see photographs of him, some old cine film of him as a younger man.

They know that he lived and breathed in ways that those other imaginary companions of childhood do not. He left documented traces, like the dinosaurs. As a verifiable historical personage, he has credentials that the tooth fairy lacks.

How do children make sense of the fact that a once-living person can stop existing in this way? The developmental psychologist Paul Bloom has argued that young children are "dualists," wired up to treat mind and body as separate entities. They get to grips with the biological facts of death relatively early (dead houseflies and ex-gerbils play their morbid parts). A person, though, is a soul as well as a body, and when the body dies, questions arise about the immaterial part. In fact, children who understand biological death perfectly well also believe in some continued psychological functioning after death. In one study, a majority of preschoolers reasoned that a dead mouse would continue to have thoughts and feelings about the events that had killed it. In a study of Spanish schoolchildren, children as old as eleven, hearing a story about the death of a grandparent, proved remarkably willing to attribute continued mental functioning after death, particularly when the narrative was framed in a religious context. The specific form of children's beliefs about the afterlife are shaped by their culture, but a readiness to believe in life after death seems pretty much built-in.

When they are confronted with it for real, death can do funny things to children's thought processes. At three, Athena found out that she was going to have a little brother or sister. When the pregnancy ended in a miscarriage, her thoughts turned to the fate of her lost grandfather. She had heard me talk about him, and seen me commemorate the anniversary of his death. She could begin to make sense of what had happened to this little not-yet person by thinking about what had happened to that poorly old man. Isaac, the happy ending to our miscarriage story, has always been more interested in the metaphysical facts. He tells us that heaven is full of dragons, and when you're in heaven "you can't get died." If the developmental psychologists are right, he is fascinated by supernatural entities and spaces because he, like all children, is programmed to think of people in terms of souls as well as bodies. In our firmly non-religious household, he has invented God for himself, rather than having the idea thrust upon him.

Telling the kids that Grandad Philip is watching over them might not, then, be so alien to their own understanding. But I realize that I am not satisfied with them simply knowing the facts about the man and loving him as an abstract entity. I want them to *remember* him: not just as a face on a photograph, but as a living, breathing, speaking person. He is part of their biographies, and I want him to figure in the stories they tell about their own lives.

I am fascinated by the psychological mechanics of what I am doing. Is it actually possible to seed a memory for someone you have never known? Can it ever become a vivid moment of experience, of the kind that can be cherished as a personal memory and endlessly relived? Strictly speaking, you should not be able to have this kind of first-person, there-in-the-moment memory for events you didn't actually live through. But it turns out that autobiographical memory is tricked in this way all the time.

The reason for this should be easy to spot. When you form a memory, you don't simply record a mental DVD of events that gets played back at the moment of remembering. Memories are constructions, made in the present moment; they are not direct lines to the events themselves. As we have seen, autobiographical memory involves close collaborations between the medial temporal lobe circuits (including the hippocampus and its adjacent cortical areas) and the control systems of the prefrontal cortex. Fragments of sensory memory are dredged up from the sensory cortices of the brain and fused with representations of more abstract knowledge about events. The entire mix is then reassembled according to the demands of the present. It is this process of active reconstruction that makes memory so susceptible to distortion.

In childhood, when the constituent systems are still maturing, autobiographical memory is even more fragile. Those first narrative constructions from early childhood can be incredibly vivid, but they are also demonstrably unreliable. They are particularly susceptible to becoming infected by nonexperienced representations of events, such as other people's stories, photographs and moving images. When you watch a home movie with your child and say, "Do you remember that day we went to the zoo?" the child is absorbing the presented image into a memory being made in that moment. Children's difficulties with monitoring the sources of information make it hard for them to tell which parts originate from the "real" memory, and which have been presented after the fact. It's inconceivable that such acts of photographically enhanced recollection will not in themselves go on to influence the child's subsequent "memories."

It seems to me quite possible that, in the editing suite of memory, my children will combine their knowledge of their grandfather's sayings with the visual images they have gathered of him, and end up with a lifelike "memory" of him speaking. Any such construction would presumably be no more nor less "real" than my sketchy memories of Dad's own parents, or than the children's memories of Dad would have been had he lived until they were four or five. In fact, twelve years on, I wonder whether my own recollections are so trustworthy. Perhaps my own memory lab has been feeding me dodgy outputs, beguiling me with constructed experiences that should be valued as a special kind of personal story, rather than as objective "truths" about how things were.

Consciously or unconsciously, accidentally or deliberately, we influence each other's memories all the time. Some of modern parenthood's greatest anxieties are about whether the children will remember the bad-tempered times, the rows and the shouting, the heavy-handed tellings-off. I have sometimes found myself wishing I could pause the tape on the kids' memory processes, on those too-frequent occasions when I have lost my temper or raised my voice. I have dreaded those future demonstrations of the tenacious power of memory. Is Athena going to turn around to me as a grown woman and recall, with devastating accuracy, the time when I left her crying in her high chair just because I wanted to listen to a particularly interesting passage of play at Trent Bridge? And what about the good times? What about the hours I held her as she fell asleep, gazing into the slowly obscured whites of her eyes, then watching her eyelids flicker open as the sounds of the house scratched at her consciousness? How I wanted to say to her, *Remember this moment. Remember how much love I had to give to you.* Remember how calm I was, how un-angry, how nonshouting. Let this be the father image you take with you through life. Whatever else you remember, remember this.

I suspect that those parental efforts to manipulate memory are successful more often than we would like to admit. We each play a part in editing our children's life stories, by talking about this and not that, by packing the camcorder on one day but not another. Children will still remember the unrecorded bits, of course, but the constant struggle for coherence that is memory-making will have one fewer prop on which to rely. I faced similar dilemmas in deciding what I was going to put into my book on Athena's development, knowing that she would probably read it one day and be confronted with one individual's partial representation of her own life. There were scenes I did not want to write about, of course, just as I instinctively stopped filming whenever she hurt herself or became upset. My book may have had the best intentions behind it, but I was still helping to determine which memories she would and wouldn't have. It felt like an awesome amount of power, and I have never felt entirely comfortable with it.

It also raised the specter of future arguments. What if she turned around to me one day and said that she disputed my account? How would we deal with the eventuality that we would remember things differently and, worse, that I had sent my version out into the world for public consumption? One friend, now in his forties, still argues with his mother about whether he once wet his bed as a seven-year-old (he remembers it; she insists that it did not happen). Does it matter that children and parents disagree about memories like this, when they doubtless disagree on so much else?

Much of how we navigate relationships must be about negotiating memories. When people get together to form families, they also have to encourage their memories to make friends. You could argue that it is one of the conditions of couples staying together that they eventually—perhaps after much disagreement—negotiate an agreed representation of events in their shared past. Those key moments in a partnership—first meeting, first kiss, moving into one's first shared home—are talked about and ultimately mostly agreed upon. It is hard to see how it could be otherwise. As Rebecca Solnit puts it, "A happy love is a single story, a disintegrating one is two or more competing, conflicting versions, and a disintegrated one lies at your feet like a shattered mirror, each shard reflecting a different story, that it was wonderful, that it was terrible, if only this had, if only that hadn't." The consensus must hold for at least as long as the relationship lasts. One friend told me that after she divorced her husband, all sorts of disagreements about memory suddenly surfaced. While they were together, disputes about details had been sacrificed for the common good.

In other adult relationships, though, it can be a different matter.

WHEN BROTHERS AND SISTERS ARE young, observed the psychologist Dorothy Rowe, they fight with each other for their parents' attention. When they are older, "siblings battle over who has the most truthful, accurate memory of their shared past." Adult siblings generally do not face the same pressures as married couples to agree on a story about their pasts. Individuals who have spent a lifetime trying to define themselves in opposition to each other are unlikely to be quite as motivated to settle their memory differences. And the fact is that adult siblings usually do not get as many opportunities as couples do to negotiate their memory disputes.

Although it was in the saddest of circumstances, just such an opportunity presented itself to Fiona and her siblings. After her father's death, she and four of her five brothers and sisters congregated at the family home to make preparations for the funeral. They could not begin in earnest until they had tracked down the sixth sibling, who was traveling in the Middle East, out of contact. In the hiatus, they spent three days sitting around the house, talking at unaccustomed length. Although the siblings were close, they were not used to spending this much time together as adults. Unsurprisingly, their conversations turned to the past. For Fiona, what followed was the perfect demonstration of how sibling memories can diverge. The five eldest children were born close together, within the space of not much more than six years. The last, Jeannie, arrived five years later. For that reason, the elder five were all able to remember the youngest child's birth and early years. And Jeannie's third birthday turned into an event that all of them could remember, albeit in their own different ways.

The family home was situated close to the Merrick, a substantial hill in Galloway in the Scottish Borders. As a birthday treat, little Jeannie wondered whether she might be allowed to climb the hill. "Jeannie may climb the Merrick," their father pronounced, "on one condition. She may be cajoled, but not carried." The little girl had approval to go up the mountain with the rest of the family, but only under her own steam. As they were growing up, the children talked about this event, and their father's memorable phrase of permission, although their memories of having had these conversations are not now particularly vivid. But when they got together for those few days after their father's death, they found that they remembered the event differently. They agreed on the basic facts of the event, but the details varied. One swore that little Jeannie's Wellington boots had been red, while others pictured them as blue. The more they talked, the more aware they became of these divergences. Rather than simply accepting their different points of view, they felt the need to make them cohere. Eventually, for reasons that no one could fully fathom, the red Wellies won out. History is written by the victors, and in the battle of memory it can be an arbitrary matter who ends up on the winning side.

It didn't always prove possible for the siblings to settle on an agreed story. Another event they all recalled was Jeannie's baptism, which was marked by the memorable detail that the baby's bootie had fallen off during the procedure. The siblings all remembered the bootie, but they disagreed about what happened afterward. Some say that there was a party at the house at which fireworks were let off. Others recall that it had been pouring with rain, which meant that fireworks would have been impossible. This time there was no détente, no recognition that one's memory was fallible and should give way to another's. The siblings kept on believing what they had originally believed, and are still living with their differences.

Why are some memories easier to negotiate than others? An obvious answer is that the people concerned are more committed to some of their memories than to others, and so are less willing to let go. But the story of Fiona and her siblings also convincingly demonstrates how two forces go head-to-head in memory. There is the drive to represent events accurately, which means being true to the often vivid impressions we have about what actually happened. And there is the drive for coherence, the need to produce a narrative whose elements fit together. In this case, coherence is a matter of agreement between people. Our stories need to make sense to us individually, but they also need to make sense to those who matter to us.

Fiona's impression is that if the memories had been more emotional, there would have been less scope for negotiation. The more something matters to us (whether it is a happy memory or a painful one), the less likely we are to change our minds about it. The siblings' extended reminiscing session also demonstrated to them how their memories were shaped by their current view of the individuals involved. On one occasion, the children had been playing on the flat roof of the house, strictly against instruction. One of them had fallen through a skylight, leaving broken glass all over the floor of the room below. Once again, the siblings agreed on the event, but not on its aftermath. It was not a transgression that could be hidden from their parents, and so they had to agree among themselves what story they were going to tell. Since David was a favorite with both parents, they agreed to pin the blame on him. To Fiona, that meant that they must have told their father first, as he would have wanted to lay down a marker for the treatment of this favored child. To others, it meant that they must have told their mother, as David could do no wrong in her eyes, and so this was the course of action that would have had the fewest repercussions.

Fiona and her siblings' memories are tied up with their feelings and beliefs about the people involved. None of their disagreements about memory were of any particular significance. They all agreed on the big stuff: how their parents treated each other; how they themselves were treated in turn. Other sibling relationships are less harmonious. In her book *The Sister Knot*, the psychologist Terri Apter describes how disagreements about childhood memories can be a source of rancor long into adulthood. Drawing on interviews with seventy-six sisters, she shows how differing accounts of the past can be challenged not just on the basis of their accuracy, but also on grounds of fairness. Feelings of loyalty and betrayal come into play. Of one sister's childhood memories, the other sister remarked: "Her memories are so twisted. Even little things, about who said what when. It's outrageous how unfair she can be."

The more emotion is invested in the memory, the fiercer the battle can be. "Our memories become part of our identity," Apter told me when I asked her about her siblings research. "If they are challenged, it's a challenge to the entire sense of who we are and how we stand in relation to other people. The person who's making a claim on my family story is telling me that I'm not who I think I am. It can be very disconcerting."

The novelist Tim Lott has thought a great deal about the long-term effects of memory disagreements. His recent novel, *Under the Same Stars*, tells the story of a pair of brothers whose failure to agree about the past has catastrophic repercussions. Lott explained to me that his relationship with his own brother is still informed by their disagreements over past events. "It's still a live issue," he told me. "If you can't trust your memory, what can you trust?" I asked him why some siblings are able to reach agreements about the past where others can't. "I think it depends on how much your identity is linked to the memory," Lott said, "but also how sure you are of your own self. If your sense of self is slightly shaky, you're more determined to hang on to your own story."

When you disagree with a sibling about your shared past, there can be the sense that the other is playing fast and loose with the cherished facts of your own life. What right does this competitor for parental affection have for rewriting your autobiography as they go? The writer Jonathan Margolis, author of several unauthorized biographies of British comedians, was well used to being accused of appropriating other people's life stories and turning them into narratives of his own. But recently the tables were turned on him in a telling way, when he found himself mentioned in the autobiography of the pop musician Louise Wener, the estranged younger sister of his wife, Sue. Several events in the book, Margolis argued in a *Guardian* article, bore no relation to his and his wife's own memories. While fully conscious of the irony of a celebrity biographer complaining about privacy violations, Margolis's piece expresses the hurt that can ensue when a sibling goes public with a divergent view of the past, and raises some important questions about who "owns" a memory.

A particular kind of memory betrayal can happen when a sibling claims for him- or herself an event that actually happened to you. A study conducted in New Zealand showed that such memories are not at all uncommon. The researchers focused on adult twins, predicting that disputes over memory ownership would be particularly common in siblings who were more likely to look similar and share personality features, as well as being of the same age and thus presumably having shared more life experiences than ordinary siblings. In their first experiment, the researchers asked their participants (twenty same-sex pairs of twins) independently to produce autobiographical memories in response to cue words. Fourteen of the pairs produced disputed memories—memories that were claimed by each twin as having happened to them alone. For example, in one monozygotic (or identical) pair, both siblings remembered going out for lunch with their mother and finding a worm in their meal. One pair of monozygotic twins seemed particularly susceptible to such errors, producing fourteen disputed memories (out of a total, for the entire sample, of thirty-six). In a second experiment, a different sample of twins were specifically asked to report disputed and nondisputed memories, and also to rate them on variables such as vividness of recollection and involvement of imagery and emotion. Intriguingly, the disputed memories were rated as being more vivid and emotionally rich than the ones on which the participants agreed, possibly because a greater effort had gone into constructing the memories that were not one's own. In a third study, the researchers found that disputed memories were also reported by non-twin sibling pairs, although the experience was not quite as common as it had been for the twins. The researchers also noted that identical twins were no more susceptible to these distortions than nonidentical ones.

A later analysis of the same data showed that there was a pattern to the claiming and giving away of memories. The researchers classified the disputed memories into those that showed the individual in a positive light (such as achievements or episodes of daring) and those that reflected more negatively (such as a memory of wrongdoing). "Self-serving" memories were more frequently claimed for the self, while those that reflected badly were more often attributed to the other sibling. If the person at the center of the memory did something admirable, or had something bad happen to them (thus qualifying them for others' sympathy), then it tended to be claimed for the self. If the star of the memory was shown in a bad light, it tended to be passed off onto the other. One fifty-four-year-old identical twin, on hearing the other claim ownership of the memory of a roller-skating injury from when they were eight or nine years old, responded indignantly, "Well, that actually happened to me if you don't mind . . . I think you'll find if you think really hard it was me." The other, yielding ground, eventually responded: "Oh well, I guess we get confused; it happened so long ago."

Disputed memories seem to be an example of how the content of our memories can be affected by the stories that other people tell. Just as a parent can instill false memories in a child, so family members can shape each other's remembering. It's not only siblings who can relate such compelling memories that we start to have them for ourselves. In recounting her memory of being a small girl and looking over a wall into her school, A. S. Byatt associates the memory with memories of her grandmother,

> a perpetually cross person, who never smiled. The year she died, she began to forget, and forgot to be irritated. She said to me, sitting by the fire at Christmas, "Do you remember all the beautiful young men in the fields?" And she smiled at me like a sensuous young girl. She may have been talking about the airmen who were billeted on her in the war—or she may have been remembering something from long before my mother was born. I shall never know. But I can see the young men in the fields.

Memories merge into memories. Byatt's grandmother's vivid remembering becomes the granddaughter's vivid imagining. Who can tell the difference? In time, we might become so convinced by other people's descriptions of their memories that we start to claim them as our own. If the experimental conditions are set up correctly, it turns out to be rather simple to give people memories for events they never actually experienced. These recollections can often be very vivid, as shown in a study by Kimberley Wade of the University of Warwick. She colluded with the parents of her student participants to get photos from the undergraduates' childhoods, and to ascertain whether certain events, such as a ride in a hot-air balloon, had ever happened. She then manipulated some of the images to show the participant's childhood face in one of these never-experienced contexts, such as the basket of a hot-air balloon in flight. Two weeks after they were shown the pictures, around half of the participants "remembered," sometimes in striking detail, their childhood balloon ride, and were surprised to learn that the photograph had not been genuine. In the realms of memory, the fact that it is vivid doesn't guarantee that it really happened.

Wade's study forms part of a tradition of research into what has become known as the *misinformation effect*. A well-known series of experiments by Elizabeth Loftus and her colleagues has shown that presenting participants with misleading information *after* they have experienced an event can change their memory of the event. For example, a neuroimaging study published in 2005 involved participants viewing, among other things, a movie of a man stealing the wallet of a girl whose neck was hurt in the process. Subsequently, some of the subjects were exposed to misinformation about the event (being told, for example, that it was the girl's arm that was hurt, not her neck). In nearly half of the cases, the misinformation was incorporated into the participants' memories of the event. The neuroimaging data showed that it was possible to predict, on the basis of patterns of activity in the hippocampus, perirhinal cortex and other brain regions, whether the false details would be incorporated into the individual's subsequent memory.

Summarizing research on the misinformation effect at that time, Loftus noted that hundreds of studies have demonstrated similar effects.

> *People have recalled nonexistent objects such as broken glass. They have been misled into remembering a yield sign as a stop sign, hammers as screwdrivers, and even something large, like a barn, that was not part of the bucolic landscape by which an automobile happened to be driving. Details have been planted into memory for simulated events that were witnessed (e.g., a filmed accident), but also into memory for real-world events such as the planting of wounded animals (that were not seen) into memory for the scene of a tragic terrorist bombing that actually had occurred in Russia a few years earlier.*

A strand of this research has concentrated on implanting "rich false memories," showing that the misinformation effect can apply to entire fictional episodes, not just details of events. People can be manipulated into remembering being lost in a shopping mall as a child, having an accident at a family wedding, or meeting Bugs Bunny at a Disney resort. Crucially, researchers know that misinformation effects in scenarios like the latter cannot stem from any genuine true memories, since Bugs Bunny is a Warner Bros. character and would never be seen at Disneyland.

The findings on rich false memories show that the misinformation effect is particularly strong when other people, especially family members, are providing the interjected information. Some benefits accrue to collaborative remembering, such as the everyday finding that couples can often help each other out by remembering bits of information that the other partner forgets. But there are negative effects as well. The term *social contagion* is used to describe the process whereby an account of an event incorporates erroneous information provided by other people. Another phenomenon, known as *collaborative inhibition*, refers to the findings that a group of people who are allowed to discuss an event actually remember less about it than the same number of people tested individually. When others are around, it seems, we are less good at retrieving the factual details of an event.

A recent neuroimaging study has provided some of the first clues to the neural mechanisms involved when our memories are shaped by other people. The Israeli and British researchers scanned the brains of thirty adults as they viewed and then (two weeks later) recalled a documentary-style film. Some of the participants were given erroneous information that they were told had been provided by other viewers of the movie. The researchers predicted that sometimes this information would actually corrupt the participants' own memories, and that at other times the subjects would simply go along with it out of a pressure to conform. The scan findings showed that persistent memory errors, which went on to become part of the subject's own retelling of the story, were associated with greater hippocampal activation (suggesting transfer to autobiographical memory) than transient errors, which seemed to be more about conforming to a public account of the events. The researchers also showed that the amygdala was particularly active when the participants thought that the information had come from other people, as compared to computer-generated representations. They suggested that the amygdala, so closely connected to the hippocampus, may play a specific role in the process by which social influences shape our memories.

One striking thing about Fiona's account of reminiscing with her siblings is how long it took for the differences in their memories to come to light. One friend, Zöe, an academic in her midforties, told me that she had only recently learned the true facts behind what she had long suspected was her earliest memory. She has an image of her father coming into the kitchen from the garage and telling her and her mother that her baby brother had just drunk some motor oil. However, her brother recently informed her that he had always believed that the event actually happened out in the garden, and the oil was intended for the lawn mower. He has his own clear memory of picking up the red can from the grass, and then the subsequent trauma of being intubated in the hospital. When Zöe discussed it with her parents, both confirmed her story, and she now thinks that she must have based her memory on conversations the family subsequently had about the event. Her brother, meanwhile, has had to accept that one aspect of his defining memory—how he came to drink the oil—might have been a fabrication. There is no doubt that he drank some oil and had the trauma of being separated from his mother and then of being intubated in a big, intimidating room in the hospital. But a key detail is disputed, and to this day they are still arguing good-humoredly over the details.

There are many ways in which we can begin to doubt memories that previously convinced us. We may discover that a sibling also claims them, and thus that we must have appropriated them from another person's life story. We may hear a differing account that makes us appreciate how much our own memory is the product of distortion, suggestion and erroneous reconstruction. Are these fake memories any different in quality from our real ones? In the first scientific study of "nonbelieved memories" (memories that people cease to believe after coming to realize that they are false), a team of researchers in the UK began with anecdotal observations that people do not stop experiencing memories as memories just because they have reason to doubt them. For example, one respondent had a vivid memory of Santa Claus climbing down the chimney, even though, for obvious reasons, she had stopped believing the memory many years before. The researchers went on to screen large numbers of undergraduate psychology students at two British universities, asking about nonbelieved memories and following up nearly a hundred students who reported having them. Those students who took part were also asked to produce a believed memory from around the same time period, along with an event that was believed to have occurred but that was not actually remembered. Participants were asked to rate the memories on various phenomenological characteristics, such as the extent to which they felt that they were really traveling back in time, and the memory's vividness, emotional intensity and importance for the self.

The first finding was that nonbelieved memories were much more frequent than the researchers had predicted. More than 20 percent of the students initially screened reported a nonbelieved memory. Based on the averages of the dates given by the participants themselves, the nonbelieved events occurred primarily in middle childhood, and participants mostly ceased to believe them in adolescence. The most common reason for ceasing to believe in the memory was that someone had told the rememberer that it was incorrect. Only a small proportion of these cases involved disputed ownership (ceasing to believe the memory because of finding out that another sibling had actually experienced the event). Other reasons given were implausibility (as in the Santa Claus case) and lack of confirmatory evidence.

The researchers then had three categories of memory to compare: believed, nonbelieved and believed but not remembered. Nonbelieved memories showed no differences from believed memories on several variables, such as visual and tactile qualities, clarity, emotional intensity and richness, coherence and mental time travel. (All of these ratings were lower for the believed-but-not-remembered events.) On other characteristics (such as auditory, smell and taste qualities, positive feelings and event significance), believed memories produced stronger ratings than either of the other two categories. As far as vividness was concerned, nonbelieved memories lay somewhere between believed memories and events that were believed but not remembered. One quality, strength of negative emotions, was particularly characteristic of nonbelieved memories, fitting with the finding from the twin studies that we are particularly likely to appropriate memories for ourselves when they show us to have suffered heroically in some way.

The researchers concluded that nonbelieved memories are similar to ordinary "true" memories in many key respects. In their study, both types of memory involved a kind of mental time travel, the re-experiencing of intense emotions and perceptual details, and the reconstruction of some of the spatial and social features of the event. Commenting on the fact that both types of memory were experienced as single, coherent episodes, the authors note that a discredited memory can nevertheless retain a compelling power.

Findings such as these confirm that we can remember things that we don't believe actually happened, and vice versa. As with my memory of the Sydney swimming pool, it is not necessary to believe that an event took place in order for us to experience it as a memory. One friend still has a clear memory of flying through the house as a child: standing at the top of the stairs, stretching out his "wings" and soaring down the staircase and through the downstairs rooms. The British study gives some clues as to why we might cease to believe in memories in certain cases. Nonbelieved memories were associated with fewer positive feelings, suggesting that representations that feel less good to the rememberer might be more easily challenged. But without research that follows the changes in belief as they happen, it is impossible to know for sure whether these differences are a cause or an effect of the challenge to the memory's veracity.

The researchers also point out that it is possible that some memories are incorrectly rejected, or "disowned," for reasons other than their truthfulness—perhaps because they don't fit with the individual's view of the self. Again, without being able to study the remembered events as they happen, it's hard to know (except in the case of extreme implausibility) whether these rejections are accurate or not. But even some impossible events are "remembered" by some of us. In addition to remembering seeing Santa Claus, a few respondents recalled seeing live dinosaurs and monsters, and having flown unaided. Whatever it is that makes a memory, it is only partly connected to the possibility that it could actually have happened.

DEALING WITH THE SLIPPERINESS OF memory is a challenge for all of us. When those memories are so central to our own sense of identity, we are naturally resistant to the idea that we could have got them wrong. But we do get them wrong, and probably more often than we think. Sometimes we accept our memories' inaccuracy, and even then continue to "remember" them. We edit our versions of the past as we go along, as our emotions change and we encounter new information, but even that sometimes isn't enough to negate the subjective power of a memory.

When I try to make sense of these sometimes counterintuitive findings, I find myself returning to the stories I tell about my dad. In my well-meaning effort to implant memories in my kids, I realize that I am taking advantage of the same special qualities of memory that the researchers have been exploring. I want my children to have what I fear losing for myself: vivid memories of my father as a living person. I want them to help me to remember him. Athena, our first child, was born two years after his death. I became a father before I had even had a chance to finish mourning him, and those family roles are probably still powerfully

entangled. Although the kids have three (thankfully healthy) grandparents, it's important to me that they should have a fourth. I want to
defend Dad against the forces of forgetting, and I am calling on the children as allies. It doesn't matter to me that their memories of him are not
strictly genuine. In this, as in other respects, they are telling their stories,
and I am giving them a helping hand.

7

THE PLAN OF WHAT MIGHT BE

HE WORKS LATE into the night, through the quiet hours after the Night Office. When the others are asleep he can meditate without distractions, as the Fathers recommend. He stands in his cell, in the dark, to spare the candle. When his legs ache from standing and pacing, he sits on the edge of his cot with his forehead resting on his conjoined thumbs. Sometimes he feels as though the plan he is holding in his head is so fragile that the slightest jolt might make it come apart. In the five months that have passed since he stood before it at St. Gall, the fabric has become eroded, as though some earthly edifice had been left out in the rain. Even on the journey back he felt that he was losing it. With every river crossed, every hitchhiked cart ride along a farmland track, he became a little less certain of the detail. He has his sketches, scratched on scraps of parchment, but the plan is too extensive for his rough efforts to encompass it. With every breath the memory becomes fainter. If Kornelimünster, the home he was returning to, had lain farther away, the image might have dissolved completely. He remembers the days spent in the St. Gall library, the hours spent trying to burn the shapes of the imaginary buildings into his mind, knowing that when he rode out of here his clouding eyes would never see them again, so long was the journey and so many the hands that needed paying.

He thinks about where to begin. The multitudes enter the *templum* through the western porch, eager to conduct their prayers and build the City of God in their hearts. So, too, through that symbolic entryway do Otgar and his brothers pass into the life of the monastery. But the knowledge stored in the huge church is so rich, it threatens to overwhelm him. So he begins outside, in the cloister. In the window of the monks' dormitory on the eastern elevation ahead of him, he sees, arranged from left to right: a black staff, a sheaf of corn, a child bearing a plate of five apples. Each object corresponds to a psalm. The building is his psalter, the repository of his knowledge. He can study, build and rebuild, combine and deconstruct at will. He imagines walking up to the child and plucking the middle apple from the plate. The words spring up to fill the space it leaves. *Lo, children are a heritage of the Lord, and the fruit of the womb is His reward.* He lets the third verse of the psalm fill his mind, reassured that his knowledge is safe. Next to the child, crawling up the walls of the monks' bathroom in the corner of the cloister, grows a luxuriant vine bearing black grapes. *Blessed is every one that feareth the Lord, that walketh in His ways.* The other psalms curl and stretch around him; he hears their words whispering in his ears. He has committed the plan of this building to memory and filled it, like an intricately sectioned storehouse, with his knowledge. He has created a library in his mind and laid away his wisdom in it, and now he can walk around it at will, consulting the treasures stored there. The work of the monk is to delight in this knowledge, to construct new thoughts in the memory of God. His pleasure and privilege are to bring this learning to mind, as he is carried by his vast love along the sacred path of meditation.

He enters the *templum* through the southeast door. After the quiet of the cloister, images crowd, some animate, pacing like beasts or flapping from perch to perch. Other objects stand passive as statues, waiting to be examined and interrogated for their meaning. He has made the images striking, unusual—a bright green lion, a man with a leopard's tail—so that they will stick better in his mind. He walks along the echoing nave, whose gilded stone ledges, effulgent in the faint light, he has inscribed with the names of the patriarchs. The altar rises in front of him, a three-columned tower depicting the Trinity. He weighs the structure in his mind against the crowding detail of the background, feeling the gravity of anxiety. He has parceled out his knowledge so that each segment can be viewed in one sweep, but in the vast space of this church the individual images accumulate, resonate, overlap. He wonders about an item or two, doubting their locations and worrying that they might have shifted with the passage of time. Forgetting is a constant fear, like the awareness of mortality in his forty-year-old bones. His knowledge of the masters—John Cassian, Chrysostom, Boethius—should be secure; these texts were beaten into him as a novice by his abbot. But everything he has learned since then is susceptible, which is why he must try so hard to maintain the elements in order.

He spends an hour or more with these wanderings. After inspecting the treasures of the main church, he passes into the separate chapel to the east, and then into the gardens and outhouses. He reviews his knowledge like a lord reviewing his property, checking for damage, delighting in the order and integrity of what he possesses. He makes Augustine talk to the psalms, watches Gregory the Great annotate the *Rhetorica ad Herennium*. The images that he has stored in this building are not mere symbols. They surprise and delight; they make the heart race and the soul swoon. His thought is a multicolored pageant of ideas behind words behind images, combining and recombining like clouds on a windy day. This is how he spends the night, in the darkness of his cell, in a steady, concerned, endlessly surprising journey, this movement of the mind toward God.

*   *   *

THE PLAN OF ST. GALL was never meant to be the blueprint for an actual physical building. The manuscript consulted by Otgar in the St. Gall monastic library, constructed of five sheets of parchment sewn together with green thread, was designed as an aid to meditation, a cognitive tool for monks embarked on the path of enlightenment. For the medievalist Mary Carruthers, the ninth-century manuscript counts as evidence that, for thinkers in the Middle Ages, memory embraced something far richer than our typical modern conception of a passive system for storing information, a mental library of static, unchanging depictions of our lives' events.

Memoria, in Carruthers's analysis, means something closer to the modern idea of "cognition." It incorporates semantic and episodic memory, thinking and reasoning, emotion and imagination. It is constructive and combinatorial, focused on building new structures rather than endlessly regurgitating old ones, and achieving this through the assembly and disassembly of many different kinds of information. It is memory as computer, not as photocopier. These are some of the qualities that fit it for the monastic task of meditation, or "the craft of making thoughts about God." To meditate is to think creatively, flexibly and generatively about spiritual perfection. In the words of a later commentator, the twelfth-century Hugh of St. Victor, meditation "delights to run freely through open space . . . touching on now these, now those connections among subjects." To make that possible, a monk needed to be able to access any part of his extensive memory of scriptural learnings. Medieval *memoria* was not intended to show off superhuman feats of learning, so much as to provide the raw data for having thoughts about the divine.

This is a thoroughly modern view of memory. As we have seen, remembering is more about recombining multiple sources of information than it is about calling to mind a fixed representation of an event. In the Middle Ages, the craft of thought was heavily reliant on the making of images, mnemotechnical devices for "seeing" one's thinking as it unfolded. Larger structures, or *pictoriae*, such as the Plan of St. Gall or scriptural descriptions of other real or imaginary architectures, provided monks with frameworks, handy blueprints for organizing their knowledge. A thinker could internalize such a plan and populate it with his own images, each of which in turn represented discrete units, or (in modern cognitive science parlance) "chunks" of knowledge. "Medieval *memoria*," Carruthers writes, "was a universal thinking machine . . . both the mill that ground the grain of one's experiences (including all that one read) into a mental flour with which one could make wholesome new bread, and also the hoist or windlass that every wise master-mason learned to make and to use in constructing new matters."

The intriguingly modern flavor of *memoria* is demonstrated in its other qualities. Memory images were selected for their emotional powers, their ability to stick in the mind and motivate thinking. Modern cognitive neuroscientists, with their interests in how neural systems of recollection and familiarity interact with the emotion matrices of the limbic system, might recognize this view of memories as inherently emotionally colored. The medieval view of memory as constructive and combinatorial fits with modern information-processing analyses that value the greater computational efficiency of constructive memory, in comparison to a data-heavy system where information is replicated faithfully in every detail. *Memoria* also specifies some of the ways in which the rememberer can combine multiple sources of information. In striving for *mneme theou*, the memory of God, a thinker would mix bits of his own knowledge and experience with representations gained from Scripture, such that the Jerusalem he imagined always contained familiar bits of landscape and architecture, and no one monk's view of spiritual perfection was quite the same as the next's.

The medieval notion of *memoria* thus gives us a new (in fact, rather old) way of thinking about the role of imagination in memory. The idea that remembering hangs on the ability to construct alternative nonreal scenarios has been a feature of writings on the topic since Aristotle onward, and is at the heart of some exciting new research in cognitive neuroscience. In fact, it may be the key to understanding why we were blessed with memory in the first place. To appreciate why, we need to ask what memory is for.

REMEMBERING EVERYTHING WOULD BE A potentially disastrous burden. In Jorge Luis Borges's short story "Funes the Memorious," the eponymous Funes suffers a riding accident that leaves him both crippled and unable to forget. "He knew the forms of the clouds in the southern sky on the morning of April 30, 1882, and he could compare them in his memory with the veins in the marbled binding of a book he had seen only once, or with the feathers of spray lifted by an oar on the Río Negro on the eve of the Battle of Quebracho." Ireneo Funes is not a particularly happy freak of nature, because he cannot see the wood for the trees. He cannot abstract invariants from the mass of constantly changing detail. He is irritated "that the 'dog' of three-fourteen in the afternoon, seen in profile, should be indicated by the same noun as the dog of three-fifteen, seen frontally. His own face in the mirror, his own hands, surprised him every time he saw them."

A feeble thinker because of his inability to abstract, Funes suffers from a memory "like a garbage heap." Total recall would be a disaster emotionally as well. In an episode of the TV medical drama *House*, a patient's family relationships are crippled because she cannot forget the past, and thus cannot forgive it. When her estranged sister tries to patch up their relationship, the patient's memory keeps returning to a long-past slight. This fictional depiction of a pathological rememberer draws on some genuine science. In a rare condition known as hyperthymestic syndrome (so rare, in fact, that only a handful of cases have ever been documented), sufferers dwell obsessively on the past and recall it in extraordinary detail. One real-life sufferer, Jill Price, has said that her mind is like a split screen, with the present unfolding on one side and the past replaying on the other. She can remember what she was doing on every day of her life since the age of fourteen. When a date flashes up on the TV, she automatically goes back to that moment in time, recalling where she was on that day and what she was doing. Surprised with the cue of October 3, 1987, for example, she replied, "That was a Saturday. Hung out at the apartment all weekend, wearing a sling—hurt my elbow."

Ordinary memories don't work like that. Our brains do not record every event in perfect detail; they don't even try. They have a better, more efficient way of connecting us to the past. They look for patterns, not details. As we have seen with memory for verbatim information, it is much more valuable to remember the meaning of what someone says than the precise words they selected. Presented with a list of words on a similar theme (such as *candy, sugar, honey, soda*), people will later claim to have seen words that weren't actually on the list but were semantically related (such as *sweet*). Our ninth-century monk would have appreciated this point. It was a commonplace in the premodern world that rote remembering was useful for certain tasks, but it was no substitute for proper thinking. As Xenophon put it, remarking on the abilities of some professional parroters, "They were very particular about the exact words of Homer, but very foolish themselves."

In fact, our ninth-century Benedictine's way of remembering demonstrates exactly the kind of cognitive nimbleness celebrated by modern-day psychologists. If Otgar had had to memorize all possible thoughts about God, it would have overloaded his cognitive resources. One strength of the recombinative model of memory is its processing efficiency. A system that would otherwise be cluttered up by extraneous detail, like the memory of poor Funes, can operate in a much more agile, data-light fashion. With damage to that system, the individual gets bogged down in details. Patients with amnesia caused by hippocampal damage make relatively fewer false-recognition errors on the word list test. They hear *sweet*, and their failure to extract the gist of the information from the word list means that they don't draw the erroneous (and quite normal) conclusion that they have seen it before. Looking at undamaged brains in the fMRI scanner, researchers see that the same basic memory network is activated when a participant gives a wrong answer (claiming to have seen a word that is actually only congruent with the gist of the list) as when giving a correct answer (recognizing a word that actually occurred). In both cases, the memory system is doing what it is supposed to do, if it is judged on its capacity to extract and retain deep information rather than act as a magnet for surface details.

In other words, our memories mostly fulfill the tasks they are charged with, skipping the details and drilling down to the real, useful meaning of the information we are trying to store. We remember what we need to remember, and forget the rest. Other memory errors, such as bias (allowing your memories to be colored by your present-day attitudes) and suggestibility (claiming memories for events you never experienced), reflect the operation of a combinatorial system that can stick together information from many different sources in re-creating an event. In the words of Daniel Schacter and Donna Rose Addis, such errors demonstrate "the healthy operation of adaptive, constructive processes supporting the ability to remember what actually happened in the past."

These findings give us a new way of thinking about the foibles of memory. It may be that memory errors are not so much failings as marks of success. Any system for information storage is fallible, but the flaws of our reconstructive memory system are more tolerable, and arguably better adapted to our evolutionary niche, than those of Funes's overliteral memory. Not only do the sins of memory shed light on the workings of our memory system, but they also provide clues to why it might have evolved. An ability to recall the past might only be a fortunate by-product of our evolution of a memory system. Its greater value, during our descent as a species, might have been its ability to foretell the future.

When we start to think about what might be involved in predicting future events, we soon see the value of storing information about what has gone. Today, for example, I am preparing for a trip to London to give a talk on memory. Although the format of the event is unfamiliar to me, I am not completely in the dark about what will happen. In order to plan for my talk, I can draw on memories of previous events I have been involved in at that venue (and many others). Although this is the first time I will have given this particular talk, I have enough experience of giving talks on this and other topics for me to have a pretty good idea about how such things go, what sorts of challenges I need to prepare for, how I can plan my timekeeping, and so on. A literal, blindly reproductive memory would be no use at all in such instances. With a memory system like those of Funes or Jill Price, I might be able to remember every detail of previous lectures and the layout of tomorrow's venue, but I would not be able to integrate that information into planning a new talk.

The idea, then, is that memory is Janus-faced, looking both to the past and the future. It is a notion that is not entirely new to cognitive psychologists. There is a long tradition behind the idea that the ultimate evolutionary function of episodic memory is to keep track of short-term goals. We evolved the ability to remember, the argument goes, so that we could keep certain important objectives in mind and then ensure that we had achieved them. In the late 1970s, the Swedish brain physiologist David Ingvar proposed that the brain simulates future scenarios concerning anticipated events, and then stores those representations so that they can later be consulted when the event in question actually happens.

One obvious implication follows from this. Whatever cognitive and neural systems are involved in memory should also be recruited when we think about what is to come. Neuroimaging research gives us clear empirical evidence that this is so. Ask participants in the fMRI scanner to imagine future events, and you see activity in similar systems to those that are activated when they think about the past. In particular, imagining the future leads to activation in the medial temporal lobe (including the hippocampus) and the medial prefrontal cortex, areas that are well established as parts of the core memory system.

If future scenarios are remembered for subsequent "consultation" in the way that Ingvar suggested, then they should be processed by the

brain like ordinary memories. Daniel Schacter and his colleagues tried to
address this question by asking participants to generate a large number
of autobiographical memories, each containing a particular individual, a
particular object and a particular place. For example, a participant might
have recalled being with a friend, Anna, in Harvard Square and having
her handbag stolen. The researchers then separated out these memory
elements and recombined them in different ways, asking the participant
to imagine future events that incorporated these new conjunctions. The
friend Anna, for example, might now be imagined drinking margaritas
at the Border Café. Replicating previous findings, greater anterior hippocampus activity was seen during the encoding of simulations that later
went on to be remembered, suggesting that future simulations are stored
by the brain in the same way as ordinary memories.

Some of the most interesting recent research has investigated the role
of emotion in the construction of future scenarios. It is well established
that the kinds of future thinking we typically do are emotionally colored. We think fondly about positive future events, and we worry about
negative ones. In fact, we have a general tendency to think about the
future in optimistic terms, showing a bias toward positive future scenarios. We also tend to forget negative events more quickly than positive
ones. Researchers in Schacter's laboratory have looked to see whether the
emotional content of future simulations affects how well they are subsequently remembered. They used the same "recombination" paradigm to
get participants to generate future events that were positive, negative or
emotionally neutral. Their findings show that positive future simulations
are remembered for longer than negative ones. In other words, the same
pattern of bias toward the positive is seen for "future" memories as for
past ones, again suggesting that future-oriented and past-oriented thinking work in similar ways.

Ingvar's "memory of the future" hypothesis also suggests that if
you have a problem in recalling the past, you should also have difficulty
in predicting the future. It was this hunch that led Demis Hassabis, a

researcher at the Wellcome Trust Centre for Neuroimaging at University College London, to test out the imaginative skills of five profoundly amnesic patients. He and his colleagues asked their participants to imagine ten new experiences on the basis of brief prompts, such as imagining that they were lying on a white sandy beach in a beautiful tropical bay, or standing in a museum and being surrounded by exhibits. Four out of five of the patients were markedly impaired on the task. Compared to healthy controls, their descriptions were less rich in terms of spatial complexity, sensory descriptions and references to thoughts, feelings and actions. To avoid the criticism that their imagination task might simply involve the rehashing of old memories (and thus put the amnesiacs at an obvious disadvantage), the researchers asked their participants to state how close their new imagined experiences were to actual memories (they weren't, on the whole). Even cueing one of the patients with props relevant to the scenario didn't help his performance. Although patients didn't differ from the controls in terms of how much they could vividly place themselves in the scenes they created, the scenes themselves were fragmentary and lacking in coherence.

The publication of these and similar findings has had an electrifying effect on the field. The leading scientific journal *Science* ranked the memory-imagination connection (including the work of Schacter, Hassabis and their colleagues) as one of the ten most important scientific breakthroughs of 2007. Particularly intriguing was the way that Hassabis's study pointed to a special role for the hippocampus in providing a spatial structure within which scenarios could be constructed. All of his amnesic patients had very specific bilateral damage to the hippocampus. Although the individual components of an episodic memory—the sensory details, for example, or the representations of specific objects—may well be stored in different parts of the brain, the amnesia study pointed to a key role for the hippocampus in binding those details together.

In fact, many researchers in this area believe that the hippocampus has two major roles in the construction of episodic memories. It is responsible for processing the associations between different features that are involved in laying down a memory in the first place. And it has a crucial second role when it comes to generating the relivable scenarios that are our autobiographical memories. It may do this by providing a spatial platform for what Hassabis and colleagues call *scene construction*. Hassabis's colleague Eleanor Maguire sees in these findings a common mechanism underlying both the recall of the past and the construction of the future, with the hippocampus "providing the spatial backdrop or context into which the details of our experiences are bound." It is no accident that the hippocampus, so important for navigation through space, is also critical for the construction of autobiographical memories. The scene construction work shows that memories are built in space as much as they are constructed in time.

THAT POINT WOULD NOT HAVE been lost on our medieval monk. Like many other animal species, *Homo sapiens* is particularly good at processing spatial information. The mnemotechnists of the Middle Ages found that they could dramatically increase their memory powers by organizing information spatially. When thinkers used images such as the Plan of St. Gall to organize their memories, they were acknowledging the superiority of this aspect of human cognition, and its usefulness as a crutch for remembering. But what those thinkers were really about was mental creation, the lifelong task of constructing new thoughts about God. Even the elaborately spatial "memory palaces" of the later Middle Ages were used to enable rememberers to think and speak creatively on the basis of their stored knowledge, rather than slavishly to reproduce bare facts.

In a sense, the spatial memory arenas generated by the hippocampus provide a neural equivalent of the medieval *pictoriae*. If Hassabis, Maguire and others are right, the hippocampus provides an imaginary space within which the rememberer can construct his or her scene. Otgar's internalized Plan implicitly acknowledged the reconstructive nature of memory. It gave him a space that he could fill with existing elements of knowledge, just as the hippocampus gives you an internal arena that you can populate with representations offered up by other parts of your brain.

The scene construction findings give us another way of understanding one of the paradoxes of autobiographical memory. For an ability that is supposed to be all about moving through time, memory isn't actually all that time sensitive. Cueing by the calendar is generally not very effective. If I asked you to recall a childhood memory triggered by the word *stroller*, the chances are that you would come up with something quite easily. If the cue I gave you was a date like *April 1975*, you might well struggle. Memory doesn't speak the language of time, and rememberers like Jill Price who can access their memories through specific dates are the exception rather than the rule. Memory is supposed to be all about our access to the past, but temporal information is not actually a very good route into it.

Needless to say, this paradox was prefigured in medieval debates about memory. In the early Middle Ages, memory was locational. Treatises such as the anonymous *Rhetorica ad Herennium* emphasized rules for the spatial organization of knowledge. With the rediscovery in the West of the texts of the classical world in around the twelfth century, a view of memory grew up as a narrative of the self (as in St. Augustine's writings), composed of a personal past, present and future. These two different views of memory—as locational or as temporal—were eventually reconciled by the scholastic philosopher Albertus Magnus, who argued that locational memory is a method for doing remembering, and gives memory its psychological *structure*, while the *content* of memory is about the past.

The modern theory of scene construction is, in some ways, a return to a pre-Aristotelian (that is, a pre-twelfth-century) view of memory. Taking time out of memory turns it into a more rootless form of imagination, and allows it to range into the future as well. Although information about the position of the self in relation to a temporal dimension is relevant to the experience, journeying into memory is actually more about traveling through an imaginative space than about zooming backward and forward along a personal timeline.

Perhaps all you need in order to do memory is a system that can generate alternative representations of reality. Some of these constructed scenes will seem relevant to *you*, and thus will come to feel owned as a memory or a bit of self-relevant future thinking. Others will come with a feeling of past-ness attached. But the temporal properties of the construction are not in themselves integral to the scene. Memory is supposed to be inherently about the past, but maybe it's not actually about time at all.

8

THE FEELING OF REMEMBERING

IN HER MEMORY, Julia is standing deep in a pine forest. It is daytime, and the filtered light illuminates narrow tree trunks stretching away on either side. Around her are scatterings of pine needles and patches of bare earth. When she looks up through the canopy of pines, she can see spiky cutouts of blue sky. There is a stream nearby, and the earth around it is rich with the smell of rotting vegetation. The stream is rocky, its dark flow whipped up into little white-water eddies. Steep banks slope down to it, and there are few signs of life. But the scene is peaceful, with the quietness of the trees and the sound of the rushing water. As the memory unfolds, Julia is there in the moment, close to the stream, inhabiting the scene and waiting for what comes next.

But nothing comes next. The narrative ends there, with this fairly typical description of an act of recollection. Julia's memory is precise and coherent, with just the sort of vivid perceptual details you would expect to see in an autobiographical memory. And yet the event never happened. She never stood in that pine forest by that narrow stream. She invented the details and placed herself among them. She constructed this scene in her imagination, and she did so because she was asked to. Julia was a participant in an fMRI study of imaginary memories conducted at University College London (UCL) by Demis Hassabis and the team behind the scene construction model. By creating memories for events that didn't happen, and allowing herself to be scanned while doing so, she and her fellow participants enabled the researchers to explore some of the neural processes that make a memory a memory.

The appeal of the scene construction model is that it reveals a common mechanism underlying both episodic memory and imagination. But memories are not the same as imaginings. Remembering does not just involve the retrieval of a more or less accurate representation of past events; it also requires that we recognize that representation *as* a memory. When Silvia's memories of her grandmother were triggered by the smell of the old lady's cigar ash, she felt very strongly that she was remembering. When you remember something, you feel that it happened to you. It involves more than mere familiarity; it involves back-there-in-the-moment recollective experience. Acts of imagination do not come with this feeling of remembering. They may feature us, their creators, as characters, but they are not tied up with our understanding of our own selves in the same way that memories are.

Scientists therefore need to understand what makes a constructed scene come to be experienced as a memory, rather than as any old imagining. They have postulated that scene construction gives us a mechanism for creating alternative representations of reality, but insist that we also need other processes to be in place to determine whether those represented events really happened or not. One approach to this question is to investigate brain mechanisms that distinguish between two kinds of scene construction: those that are experienced as memories, and those that are not. The study in which Julia participated was intended to extend the neuroscientific model of Hassabis and colleagues so that it was better able to account for the specifics of episodic memory.

With this aim in mind, the researchers asked their volunteers to construct three kinds of scene in the scanner. They were asked to generate real episodic memories, and they were asked to produce memories for an imagined event (such as standing by a stream in a forest) that they had created in a prescan interview a week earlier. They were also asked to create some brand-new fictitious scenes: novel imaginary memories. As the scene construction model would predict, the real and imaginary memories looked similar in terms of the brain activation they produced, with a thorough recruitment of the core memory system. But the researchers' design also allowed them to explore differences between the two kinds of memory. Compared to imaginary memories, real memories were characterized by activation in three specific brain areas: the anterior medial prefrontal cortex, the posterior cingulate and the nearby precuneus. These are all regions that lie outside the core memory system of the medial temporal lobe.

These fMRI findings fit with what is already known about the functions of these brain areas. The anterior medial prefrontal cortex (sited an inch or so behind your forehead) and posterior cingulate are thought to be particularly involved in self-reflection and thinking about other minds. In the UCL study, the precuneus turned out to be especially active in memories for previously imagined events (compared to brand-new imaginary memories), fitting with evidence from other studies that this part of the brain has a role to play in making constructed scenes seem familiar. The scene construction system generates an alternative representation of reality, and the precuneus "tags" it as a scene that has actually been experienced before.

Julia's imagined recollections also fit in with previous research suggesting that similar processes underlie the generation of real and imaginary memories. When scientists used the technique known as EEG to look at patterns of cortical activity during performance on imaginary memory tasks, there was evidence of particularly high activity in the left prefrontal cortex, consistent with the idea that this is where some of the effortful stitching-together of memories takes place. In other respects, cortical activations look quite similar between real and imaginary memories. Whatever basis we use for deciding whether something really happened or was merely imagined, it must be a complex decision process based on many different sources of information.

All of this research adds to the developing picture of how autobiographical memories are created. Presented with certain cues to retrieval, the core network generates a representation of an event that includes the basic spatial context along with relevant objects and characters in their appropriate locations. For example, if you are remembering the time when your handbag was stolen in Harvard Square, the hippocampal system constructs a representation of the physical context (Harvard Square) and populates it with the other key features: your friend Anna, the handbag, and so on. In the case of an actual memory, this process of scene construction is accompanied by activity in the "add-on" brain regions that support feelings of familiarity, trueness and relevance to the self. This extra activity is the basis for our knowing the difference between an event we actually experienced and one that we (for whatever reason) imagined. Together, the add-ons combine to produce the characteristic "feeling of remembering" that allows us to own the memory as something that actually happened to us. As with any complex system, this intricate interplay of psychological functions can sometimes go wrong.

I "REMEMBERED" GOING TO THE swimming pool on Cremorne Point because I had so strenuously imagined it. When I lost my ability to adjudicate between memory and fantasy, I plumped for memory. In a sense, I suppose, I *was* remembering: I had a memory for something I had imagined. The imagining was there, and the feeling of remembering was there. Together, they were enough to persuade me that the event actually happened.

This particular kind of false memory has now been investigated quite thoroughly by experimental psychologists. Several studies have shown that the process of imagining an event makes people more likely subsequently to have a memory for it. This conversion of an imagining into a memory is known as *imagination inflation*. In one experiment, volunteers went on a walk around a university campus with a researcher and were asked either to perform or to imagine performing some weird and not-so-weird actions. For example, at a campus Pepsi machine, the relevant action was either to check the machine for change or to get down on one knee and propose marriage to it. For both familiar and bizarre actions, the mere act of imagining performing the action led to subsequent false memories. You only need to imagine proposing to a Pepsi machine once to run the risk of having a false memory for having done so.

The phenomenon of imagination inflation has serious implications for the way we treat memories of childhood. In one study, researchers asked adult participants in the UK about a series of events from their childhoods and asked them to imagine them happening. One of the events (having a milk tooth extracted by a dentist) was a familiar one judged likely to have happened to many of the participants. The other event (having a nurse remove a skin sample from a little finger) was chosen to be implausible (in fact, the researchers went to the trouble of checking medical records to confirm that this procedure had never been conducted in the UK). Participants who had been asked to imagine the "skin" event were four times more likely to remember it happening to them than participants who had only been exposed to information about it. In addition to generating false memories, imagining events also made people more certain that the events had happened to them. The researchers concluded that merely imagining an event from childhood can produce false memories of it.

Imagination inflation also shows its effects in future-oriented thinking. Karl Szpunar and Daniel Schacter recently asked participants to simulate future scenarios involving personally relevant features, in a version of their "recombination" paradigm. For example, a volunteer might be asked to imagine being lent a dollar bill by a friend, James, in order to buy coffee from Starbucks. The next day, participants were asked to re-simulate the same scenarios, but differing numbers of times. Some par-ticipants produced one simulation, while others did it four times in total. They were then asked to rate how plausible the scenarios were. The results showed that repeating the simulation made the scenario seem more plau-sible, although this only held for positive and negative scenarios, not neu-tral ones. If a future event seems far-fetched, you might conclude, all you need to do is think more about it. The more imaginative effort goes into creating a future scenario, the more possible it begins to seem.

Memory researchers have observed that advertisers make deft use of this phenomenon. To test this idea scientifically, researchers manipulated the vividness of imagery in advertisements for a fictional new popcorn product. A week after participants had viewed these advertisements (and some had actually gotten to try the product), they were tested on their memory. Those who had seen the high-imagery advertisements were as likely to say that they had tried the popcorn as those who actually did. These false memories were highly confident, and they were also asso-ciated with favorable ratings of the product. The popcorn wasn't just remembered; it was remembered fondly. This "false experience" effect, as it has been termed, points to one way in which high-imagery advertising can have its effect. Certain ads don't just succeed in making us feel we want to try something; they trick us into thinking that we have already done so.

The precise mechanism through which imagination influences mem-ory is still debated. One explanation is that people make erroneous judg-ments about what is real and what is not, based on information about familiarity and sensory detail. When a scene is rich in sensory and per-ceptual detail, it is more likely to be judged as having really happened. The act of imagining is likely to increase the amount of detail in the rep-resentation, as was undoubtedly the case in my memory of the Sydney swimming pool. I thought so long and hard about the details of the scene, it became as real (and probably more real) in my imagination as any true memory.

But the act of imagining also involves a lot of cognitive effort that you don't find in ordinary remembering. My "memory" of the swimming pool was hard-won, and if I had been able to use information about how hard I had sweated, imaginatively, in judging whether the event had actually happened or not, I might not have made the error. According to one influential model of memory, I would usually be able to do just that. For around four decades, the psychologist Marcia Johnson and her colleagues at Yale University have been looking at how people distinguish between the sources of different kinds of information. This detailed body of research has led to a model of memory, the *source monitoring framework*, that carries some surprising implications for our thinking about memory. It is well established that a memory is not a faithful, unchanging representation of an event. Instead, the source monitoring researchers tell us, it is an attribution or interpretation that we make about a particular mental experience. An event plays itself out in our heads, and we decide, on the basis of many different kinds of information, whether it actually happened to us or not.

We go about making these attributions by weighing up various bits of evidence associated with the thing we are experiencing. Subjectively speaking, memories for events that have been imagined are not as perceptually rich and contextually detailed as real memories. We can therefore use differences in the detail of a memory to help us decide whether it happened or not. We also use information about the amount of mental effort that was involved in generating the mental experience. If it comes to us easily, we assume that the event in question was really experienced.

The problem is that these bits of information are not always easily distinguishable. A particularly vivid imagining (containing all the perceptual detail that would typically indicate a real memory) could easily be taken as having actually happened. We have already seen how disputed memories (when two siblings each claim a remembered event for themselves) are rated as more vivid and emotionally rich than nondisputed ones, suggesting that the mistaken sibling spends so much mental effort imagining the other's memory that they end up believing it actually happened to them. In the case of memory disputes between siblings, it only takes an occasion for the siblings to start reminiscing together for it to become quickly apparent that someone has misremembered. But in other cases of vivid imagining, where the rememberer's testimony may go unchallenged, the falseness of a memory may not come to light so readily.

The evidence from neuroimaging, too, points to true and false memories as having fairly similar neural signatures. One study of the misinformation effect found that true and false memories produced differing patterns of activation, but that those differences only obtained in "early" parts of the sensory system that may not have been accessible to consciousness. Jon Simons and his colleagues at the University of Cambridge have proposed that activity in the anterior medial prefrontal cortex—that little patch of brain right behind the center of your forehead—has a particular role in distinguishing mental contents that were internally generated from those based on genuine perceptions. This region seems to play a major role in "deciding" whether a particular mental experience was self-generated or came to us from outside. If the data feeding into that judgment system looks the same for true and false memories, then it will be a considerable challenge for that neural system to make accurate judgments.

Not only are the data noisy, but our judgments are also affected by various biases. We use different criteria on different occasions, depending on how the question is put to us, what is at stake, and what we want from the answer. If we are under pressure of time, for example, we can more easily jump to the wrong conclusion. Asked by a friend to share a memory over a glass of wine, we might be blusteringly adamant that the event really happened. If we are testifying in court, where other people's futures could be at stake, it might be a different matter.

The source monitoring framework thus gives us a useful model for understanding the foibles of memory. Like the scene construction model, it holds that true and false memories arise in our minds through the same basic processes, and it provides psychological detail about how we can get confused between them. It explains the extensive evidence that false memories can be implanted in our minds—through the encouraging of imagination, through the feeding in of misleading but irrelevant information, or through pulling on the strings that can manipulate our reality judgments. The net effect of source monitoring errors is to make certain self-generated mental events come to be taken as real memories. The information on which we base those judgments is not perfect, and neither are the processes involved.

There are plenty of reasons, then, why making judgments about the source of a mental experience is a complicated business. It involves making subtle distinctions among kinds of information that are often not readily discriminable. One of the most tricky factors is emotion. Johnson and colleagues' experimental work shows that emotion is often used as a basis for making a reality judgment ("I know it happened because I feel it so strongly"). But emotion is also known to reduce the accuracy of source monitoring judgments. When people are focusing on the feelings involved in an event, they pay less attention to the perceptual and cognitive information that might allow them to make an accurate reality call.

My swimming pool "memory" demonstrates the fallibility of these judgments. A huge amount of cognitive work was involved in creating the mental experience, and this should have nudged me toward accepting its falsehood. But other factors were working in the opposite direction, such as my repeated efforts at vivid, perceptually rich imagining, at the same time as I obsessed about the emotions of the event. Nonbelieved memories, such as those subsequently disbelieved memories of having seen Santa Claus, must be another example of imaginings that somehow take on sufficient perceptual force to trigger an attribution of them actually having happened. One colleague told me that she had a friend who had a false memory of having an arm amputated. Perhaps, for whatever reason, this woman had been anxious about losing her arm, and had imagined the operation in gruesome detail. That imaginative process furnished enough sensory and perceptual detail for it subsequently to have the force of a memory.

This leaves us with a problem of how we can ever decide whether a memory is true or not. In the last couple of decades, this question has assumed huge proportions in the legal system. In fact, the "Memory Wars" of the early 1990s, in which memories of child abuse "recovered" during therapy had their authenticity challenged in highly publicized debates and lawsuits, have played a big part in alerting scientists and the general public to the constructed nature of memory. Such "recovered" memories are now known to be unreliable. A study published in 2007 compared memories of childhood sexual abuse produced by three different groups: those who had always remembered that they had been abused; those who had suddenly remembered abuse outside therapy; and those whose memories had been "recovered" through therapy. Compared to the continuous memories and those that had suddenly recurred outside therapy, the "recovered" memories were less likely to be corroborated by other people or other sources of evidence. In fact, no corroborating evidence was found for any of the abuse memories in the "recovered" memory group. The researchers concluded that some of the techniques used in recovered memory therapy, such as visualization, suggestion and hypnosis, could have been responsible for these differences. As we know from the research on imagination inflation, asking people to vividly imagine abuse events is likely to lead to subsequent false claims that the events actually happened.

The researchers further supported this conclusion by asking how surprising the abuse memories were to the individual. "Suddenly remembered" abuse memories were more surprising to the rememberer than memories that had been recovered during therapy. This was interpreted as showing that prior expectations about the purpose or outcomes of the therapy had contributed to the generation of false memories. Other experimental findings have shown that people who claim memory recovery through suggestive therapy are more likely, in a simple lab test of false memory formation, to claim memories for events that have not happened to them. This suggests that such individuals may have a natural bias toward false memories that can, in some circumstances, be capitalized on by unscrupulous therapists.

"The difference between false memories and true ones," observed the legendary surrealist painter Salvador Dalí, "is the same as for jewels: it is always the false ones that look the most real, the most brilliant." So contentious is the issue of false memories that the British Psychological Society recently released a set of guidelines on "Memory and the Law," aimed at ensuring that legal practitioners can evaluate the testimony of witnesses on the basis of the best available scientific evidence. Faced with overwhelming evidence for the fallibility of memory, lawyers and judges are now much more cautious about relying on uncorroborated eyewitness testimony or unsubstantiated accusations of abuse.

Researchers have recently tried to apply findings from laboratory studies of memory to determine whether reported memories are true or false. One particular feature of memory reports, the amount of detail supplied, has had some success at distinguishing true from false memories, although it is not yet considered reliable enough for use in legal cases. In one diary study, two volunteers were asked to record events that had actually happened and some that they had made up. When reading them back seven months later, the true memories came with a feeling of recollective experience (taking the participant back, to some extent, to that surreal in time), while the false memories tended to come only with feelings of familiarity. In another recent study, researchers analyzed narratives produced by adult participants about real or imagined childhood events. They found that narratives of real events were longer and contained more reference to cognitive processes (for example, beliefs and knowledge), while imagined events contained more reference to emotions. Despite the presence of objective differences between false and true memory reports, several studies have found that people asked to act as raters cannot, in practice, reliably distinguish between testimonies about real and imagined events.

Source confusion errors can also work the other way around. It is presumably possible to have memories from which the "this is real" tag gets detached. For example, if the add-ons identified by Hassabis and colleagues stopped working in sync with the scene construction system, the tag that tells the individual "this is a memory" might get removed. Another way of looking at this is to say that, for some individuals with very vivid imaginations, intentions and thoughts about the future might look very much like memories. The source monitoring task of deciding what is a real memory would then become that much more difficult.

There is evidence that this is exactly what happens in some psychiatric patients. It is well established that patients with disorders such as schizophrenia, where activity in the prefrontal cortex is often disrupted, can have particular difficulties with source monitoring. Confusion between the real and the imagined might lead some such patients to see nefarious significance in stimuli that are actually part of their memories. The clinical psychologist Tony Morrison of the University of Manchester has described the case of Joe, a psychotic patient who was paranoid about being followed by assailants in a white Transit van. He had a vivid, recurring image of being bundled into the back of such a van and being repeatedly attacked with a variety of weapons. Through cognitive therapy, it transpired that this image of the attack bore some relation to an assault he had previously experienced in prison, while also incorporating features of a separate scene he had witnessed on television. Joe's fear of white vans was a memory repackaged as a premonition. He thought he was foreseeing a horrific event; he did not realize that he was actually remembering one.

It is not just psychiatric patients who suffer from source monitoring errors from time to time. We all get confused on occasion between what we have imagined and what has actually happened. Hatching a plan to go to the supermarket later today, I formed an intention to write certain items on the shopping list. Unable to remember clearly whether I had written them down or not, I just had to go back into the kitchen to check. (I had.) Researchers have begun to uncover the neurological bases of these abilities. A study from Jon Simons's lab at Cambridge recently showed that the presence of a particular fold in the anterior medial prefrontal cortex, known as the paracingulate sulcus, relates to people's performance on experimental tests of reality monitoring. Participants with the fold were better able to distinguish between an event that had been imagined and one that had actually occurred, compared to those who lacked the fold.

Dreams are one kind of mental event whose provenance can be particularly difficult to determine. At least some of the things that we think we remember might actually be things that we dreamed. Dreams are a special case because they are of an unambiguously internal origin, but also because they would not figure in our waking lives at all without memory. If you cannot remember a dream long enough to report it when you wake up, you can never know about it. If you do remember it, a simple error in source monitoring can lead you astray about its correct attribution. One friend told me that she went through a phase of having rather boring dreams, such as the milkman coming to the door asking for his money and my friend paying him. But one time a source monitoring error led her to confuse her dream with the actuality. When the milkman came for real to the door, she insisted indignantly that she had already paid him. She hadn't, of course; she had only dreamed it. The error soon came to light, and the milkman went away with his bill happily paid.

In other cases, the problem is not so easily resolved. In patients with certain kinds of brain damage, the line between reality and imagination comes under more sustained attack. When that happens, people start to tell stories.

9

REMEMBER ME A STORY

A NURSE IN her early fifties, Claire greets you with a smile of such casual familiarity that you could be forgiven for wondering whether you have met her before. Her fine dark hair is cut in a bob, and the anxious, kindly eyes behind her thin-framed glasses give her the concerned look of a friendly schoolteacher. Before I meet her for the first time, on a breezy, sunny September day in Cambridgeshire, I am told to wear something distinctive. I choose a badge that says READ HENRY MILLER in bold black type, celebrating a literary obsession of mine from my early twenties. If I make sure my appearance is distinctive, Claire will have something other than the details of my face on which to hang an identity for me. Even her husband, Ed, does it. He wears a shark's-tooth necklace wherever he goes, so that Claire will see him and realize that it's him, the guy with the shark's-tooth necklace, her husband. He does it for that very simple reason: so that she will know who he is.

Claire suffers from profound anterograde amnesia, which means that she cannot convert short-term memories into new episodic memories.

But her amnesia is also retrograde, meaning that she has lost memory for events that happened before the illness that damaged her brain. Her memory problems look both ways. If you ask her what she remembers, she will say that she has lost all memory of her life from her late teens to when her illness struck in her early forties. Particularly distressing is the fact that she cannot remember any episodic detail relating to the lives of her four children as they grew up. Her memory for her adolescence and earlier childhood is also patchy. Describing her problems, her family likens her memory to a "locked vault," with the few memories that she can relate having the suspiciously slick quality of well-rehearsed stories.

Claire understands the extent of the neurological damage behind these problems. At the age of forty-three, a common virus known as herpes simplex attacked her brain on both sides, causing most of its damage in the right hemisphere. Brain scans showed extensive lesions to the memory circuits of her right medial temporal lobe, with some frontal damage as well. In addition to wiping out her memory, the virus also destroyed her senses of smell and taste. Claire's understanding of her own condition is quite sophisticated. She knows that she doesn't remember. She also knows that her problems are limited to episodic memory. Her semantic and procedural memories are fine: she knows the names of her four children, and the capitals of Europe, and she is perfectly capable of remembering how to operate her computer or bake her own bread. In this respect she has much in common with the most famous amnesic patient in history, Henry Molaison, who died in 2008 after fifty-five years of being unable to create new memories. Molaison had undergone an experimental procedure to treat chronic seizures, which had left his hippocampi seriously damaged on both sides. For reasons of patient confidentiality, for most of his life he was known to the world by the initials HM. Like Claire, HM was unable to recognize family members and scientists with whom he had worked for decades. In a 2007 interview, one of the neuropsychologists who studied him, Suzanne Corkin, observed: "He is in my PhD thesis and I have followed his progress for the last forty-three years. And he still doesn't know who I am."

The most convincing demonstration that HM had a specific problem with episodic memory came when his neuropsychologist, Brenda Milner, asked him to draw a complex pattern in a mirror. Copying a drawing without seeing your own hand, only your hand's reflection, takes some tricky adjustments to your usual perception-action linkages (as you will know if you have ever tried to trim nasal hairs in a bathroom mirror). HM eventually mastered this difficult copying task, even though he could not recall individual occasions on which he had been trained in it. Like Claire, his procedural memory was intact—he could still help out in the garden or fix himself some lunch—but he could not lay down new episodic memories.

I am wearing the READ HENRY MILLER badge because Claire also has a particular problem with recognizing faces. Damage to her fusiform gyrus, toward the back and bottom of the temporal lobe, has left her with a condition known as prosopagnosia, or face blindness. She can literally see my face, of course, but she will not remember it. In the first days of her illness, the inability to recognize people who should have been fully familiar to her was incredibly distressing. Thinking about that confusion now still upsets her; she remembers the feelings if not the details of the events, in particular her frustration that she could not function in the way she used to: as a mother, wife and professional nurse. She had lost all feeling of connection to the people who mattered to her. "When you meet an old friend," she tells me, "you want to ask them about their lives, about their children. I couldn't do any of that. They weren't getting what they wanted from me. I wasn't giving them back what they needed."

I have been brought to meet Claire by Catherine Loveday, a neuropsychologist from the University of Westminster. Catherine's research team first heard of Claire's case when two other neuropsychologists, Narinder Kapur and Barbara Wilson of the University of Cambridge, passed on her details to them. Kapur, Wilson and their team had been exploring a new piece of technology, under development by Microsoft Research Cambridge, as a possible therapeutic tool for use with amnesia patients. Called the SenseCam, this new gadget is a small digital camera about the size of a box of Café Crème cigars, designed to be light enough in weight to be hung around the neck by a lanyard. A fish-eye lens captures wide-angle images of the visual scene, which are stored on a flash drive for later transfer to a home computer. Software allows the user to review the images at their own pace, at a time that suits them. The camera takes a picture at a fixed interval, such as thirty seconds, but also incorporates sensors for detecting movement and changes in light and temperature. If anything interesting changes in the visual field, SenseCam is triggered to record it. The device is designed to be low-powered, so that it can be left on all day without the need for frequent recharging. It also contains an accelerometer, which allows the device to stabilize the image if the camera is shaking at the time when a picture is triggered.

SenseCam was initially developed as a technology for keeping a visual diary of the user's day, and was quickly adopted by fans of "lifelogging," the growing movement toward digital archiving of the tiniest details of our lives. But Microsoft Research soon recognized SenseCam's possible application for treating memory disorders. For years neuropsychologists have promoted the use of memory aids to help amnesiacs with some of their day-to-day difficulties. Patients are encouraged to use calendars and mobile phone alerts to remind them to take medication or keep appointments. But these strategies are mostly focused on prospective memory: remembering to do things in the future. Until recently, memory aids have been of much less use in helping dense amnesiacs to keep some footing in their own pasts.

SenseCam provided an exciting new alternative to these traditional therapeutic aids. In collaboration with Kapur and Wilson at the Memory Clinic at Addenbrooke's Hospital in Cambridge, scientists at Microsoft Research began a trial with a sixty-three-year-old patient known as Mrs. B, whose hippocampi had been damaged by a condition known as limbic encephalitis. For eleven months, Mrs. B wore a SenseCam to record interesting or out-of-the-ordinary events in her life. A day after each event, the patient's husband asked her if she recalled the happenings of the previous day. They then reviewed the same SenseCam images together on a total of seven occasions. When the SenseCam phase of the study was complete, Mrs. B and her husband undertook a control phase in which events were recorded in a diary. The results showed a clear benefit to reviewing the images. In the SenseCam condition, Mrs. B remembered around 80 percent of recent, personally experienced events (compared with 49 percent in the diary condition). She also showed continued retention of the SenseCam events up to three months after she had last viewed the images (the events themselves had occurred up to eleven months previously). Retention was much poorer in the diary condition, and keeping a diary turned out to be so laborious and ineffective that this phase of the study was brought to an early conclusion.

What explains these persuasive findings? One obvious point is that SenseCam capitalizes on the highly visual nature of autobiographical memory. The fact that Mrs. B was recalling the events themselves, rather than remembering her viewing of the SenseCam images, is confirmed by Mr. B's reports that she would often recall features of the event that weren't depicted in the images. That said, the wide-angle field of view of the SenseCam images means that the pictures are already rather rich in detail, if not in visual quality. They thus incorporate the kind of trivia that we might expect would be good cues for memory. SenseCam images are also taken from the perspective of the self as it is experiencing them, which arguably brings them closer in nature to visual autobiographical memories, and may help them to be consolidated better in the medial temporal lobe system.

Researchers have now started to test out these ideas by asking how viewing SenseCam images relates to changing activations in the brain. One set of findings compared activation in the medial temporal lobe memory network in two conditions: the remembering of events recorded in a diary and the recalling of events for which SenseCam images had been viewed. Activation of memory areas was far greater for the SenseCam events. In addition to the memory circuits, there was particularly strong activity in the visual areas at the back of the brain, where we know that sensory-perceptual fragments of autobiographical memory are stored. In the words of one of the researchers, Martin Conway, these are the same kind of activations you would see if the person lying in the scanner was actually viewing the events for real. Judging from the neuroscientific evidence (while being cautious about assuming that more intense neural activity actually translates into a more intense subjective experience), it seems plausible to conclude that SenseCam creates a more vivid experience of remembering.

In ongoing studies, Conway, Loveday and colleagues are scanning Claire's brain as she recalls events for which she has previously viewed SenseCam images. Preliminary findings suggest that Claire's brain shows an increase in activation in visual areas, compared to a condition where she recalls events that she has recorded in a diary. In other words, Claire seems to show the same pattern of activations in response to SenseCam images as the participants with intact brains. Conway and Loveday are hopeful that these findings demonstrate a neurological basis to the positive effects of SenseCam seen in Mrs. B and other amnesiacs, and may help us to understand exactly what aspects of Claire's remaining brain function are most critical for her remembering.

IT SEEMS CLEAR THAT SENSECAM changes the experience of remembering in some way. To find out more, I went for a walk in Leeds, a city I have only visited a couple of times. My companion for the day, the neuropsychologist Chris Moulin, suggested that I try SenseCam out by going for a walk in an unfamiliar place, and later reviewing the images to get the full experience. So we set off from the university campus on a hot, sunny July day, the little black box bumping around on my chest. As we walked, I tried to forget about the experiment and focus instead on a conversation with Chris about memory. When I got home that evening, I uploaded the images onto my computer and intentionally forgot about them for a month. I wanted a decent stretch of time to pass before I looked back at them. It was hardly scientific, but a month seemed to leave enough time for ordinary forgetting to do its work.

When it came to looking back at the images from my walk, I felt my usual concern for the sanctity of the art of remembering. I made sure I was alone in the house, and able to concentrate and follow my memories wherever they led me. Chris had told me that viewing SenseCam images is not like looking at ordinary photographs. You remember more stuff, he said: thoughts, conversations, feelings. I wanted to be prepared to catch anything that might flash into my mind. Before I reviewed the images, I had some general impressions of the walk. I remembered the sunshine, and some of the details of the conversation, and the fact that Chris had taken me to the Dark Arches area of Leeds where you walk along underneath the railway. But if SenseCam was all it was cracked up to be, I would remember much more than this. What other forgotten moments would come to mind?

I began by creating a time-lapse movie from the thousand or so images that the SenseCam had stored. I set the frame rate to five frames per second—quite fast, but not so fast that I couldn't process separate images. I decided that I would stop the playback if I wanted to focus on a particular image. The movie began with the scene in Chris's office, and then suddenly switched to the corridor outside. The sight of the greenish walls lit by fluorescent tubes triggered something, and I had the impression of a new memory coming into consciousness. I paused the playback, recalling how I had gone to the loo and then waited outside in the corridor for Chris while he checked his pigeonhole for mail. That was not in my memory before I started watching the movie. SenseCam had its first success.

I restarted the playback. Suddenly we were outside, in the sunshine on the streets of Leeds, walking down Woodhouse Lane to the city center. I was able to pinpoint certain details of our conversation to points along the way: the square where we had discussed Loftus's work on memory erasure, the bus stop where we had talked about memory for surgery under anesthesia. A long sequence then followed in which the camera recorded my perspective across the table from Chris as we had a lunch of Japanese food. I remembered exchanging stories about our weddings, both of which were rather unconventional. I could almost hear our conversation in my head. I saw my own hands flapping around as I grappled with chopsticks, and shamefully noted a few occasions when I sneaked a glance at my phone under the table. After lunch we walked along the river and into the shopping district. I had a vivid memory of a text coming through on my phone, saying that my daughter, Athena (who had been playing a cricket match that day), had just taken a five-wicket haul. I saw myself saying good-bye to Chris and walking back to the university on my own, then getting lost and needing to use my satnav to find my way back to where I had parked my car. I must have forgotten to switch the camera off, because the rest of the movie showed the ceiling of my car, from the perspective of the SenseCam laid faceup on the passenger seat, faithfully recording the changing patterns of light and the summer foliage flashing past the window.

It seemed that SenseCam's power had been demonstrated to me. I had no way of making scientific comparisons, but I felt that (as Chris had predicted) I had recalled more than I would have if I had only been looking at photographs. When you take a photograph, you are intentionally bracketing off a moment of experience, selecting what you want to include in the representation and what you don't. SenseCam is much more indiscriminate. It notices things you didn't notice. In one study of the effects of viewing SenseCam images, participants reported a sense of wonder at how the technology had made their own lives look so strange to them. One participant reported: "I took it on holiday and 80 percent of the photos were of my boyfriend . . . but what I loved about it was the way it caught his mannerisms and behavior . . . the way he'd be looking out the window or watching something else."

Participants also reported how SenseCam was able to "foreground" moments that would otherwise have passed unnoticed, just like a good novel or movie would. Sometimes this was a cause for reflection, as when participants commented on how much of their lives was taken up with mundane activities like sitting in a car or washing up, and sometimes it was even an engine for personal change. I certainly learned something about myself by reviewing the events of my walk in Leeds, such as how much I gesticulate with my hands when I am talking. When I looked through the images more slowly, flipping through them with the cursor key at my own pace, I noticed even more details. I saw students dressed up for graduation, which reminded me of my feeling of surprise that such momentous events were happening to a group of people of whom I was only dimly conscious as the day had begun: surprise that it was happening, and surprise that I had not noticed. I saw an image of Chris giving directions to the driver of a van, an event I had completely forgotten. I knew that we had been stopped for directions on another occasion that day, but when I later saw the image of the young Asian man who had stopped us, and saw his forlorn expression, I remembered my disappointment at myself that we had not been able to do more to help him find his way. The images were grainy, and distorted by the fish-eye lens, but they had an uncanny ability to bring the past to life and to make me experience it in a different way.

Indeed, Martin Conway has argued that reviewing such images can lead to what he calls *Proustian moments*, in which a tiny detail such as a gesture or detail of decor can trigger a flood of remembering (just as Marcel's taste of the *petite madeleine* led to the cascade of recollections in his great novel). Leaving aside the objection that the term *Proustian moment* does not quite coincide with the effortful process of recollection that was actually described by Marcel, these sudden, vivid experiences of recollection are undoubtedly a part of our remembering experience, and might conceivably be triggered by viewing SenseCam images as strongly as they are triggered by smells or music.

Conway's explanation of this phenomenon forms part of his broader and highly influential theory of autobiographical memory. At the center of his work Conway has placed the idea of *autobiographical knowledge*, a kind of semantic memory for how the events of our lives have unfolded. For example, I know certain facts about my own life (such as that I was a pupil at Kingswood Primary School in Basildon between 1975 and 1978) that don't necessarily have autonoetic, episodic memories attached. Any particular events that I remember from this time are framed by that conceptual knowledge about how my life has unfolded. In fact, an autobiographical memory happens when fragmentary, perceptual episodic information stored in the sensory cortices becomes connected to autobiographical knowledge structures, so that the memory can acquire a personal dimension and the remembering self is placed conceptually and experientially in time. Autobiographical knowledge provides the skeleton that gives structure to our memories. This crucial integration of episodic images with autobiographical knowledge also explains why we can access episodic memories through autobiographical knowledge cues, such as I would do, for example, if you asked me to recall an event from my primary school years.

Conway's theory is thus an example of a reconstructive theory of memory that views memories as constructions built up from multiple sources of information stored in different neural systems. But the theory also needs a way of explaining why memories are experienced as having happened *to us*. Conway has proposed that there are certain "cognitive feelings" that signal to the experiencer what cognitive state they are in. For example, the "feeling of remembering" that accompanies the outputs of your memory system tags the experience as a memory, rather than as a dream or hallucination. We saw in the last chapter how this feeling of remembering is thought to have its neural roots in the add-on regions of the scene construction system, such as the precuneus, the posterior cingulate and the anterior medial prefrontal cortex. If you thought back over the events of yesterday and then tried to insert a memory for something that hadn't actually happened, it would stand out. It would not feel genuine, because you would not experience that authenticating feeling of remembering. In normal circumstances, the feeling of remembering guarantees that the memory feels as though it happened to us. In other situations, there can be a mismatch between the contents of consciousness (the actual memory) and the associated cognitive feeling. When that slippage happens, remembering can go awry in some distinctive ways.

Chris Moulin, my companion that day in Leeds and loaner of the SenseCam, has spent much of his career studying one particular class of these memory anomalies. As a doctoral student at Bristol, he received a letter from a GP describing an eighty-year-old former engineer, an immigrant to the UK from Poland, who had been complaining of memory problems. The doctor had suggested to the patient, who became known as AKP, that he attend a memory clinic; the patient's response was that there was no point in going to that clinic because he had already been. His sense of having experienced events before was practically constant, and exacerbated when he encountered novel stimuli. He would not read newspapers or watch TV because he had seen the articles and the programs before. "However," Moulin and his colleagues wrote in their description of the case, "AKP remained insightful about his difficulties: when he said he had seen a program before and his wife asked him what happened next, he replied, 'How should I know, I have a memory problem!' "

The anomalous feeling of having experienced a moment before is of course not necessarily a sign of pathology. In his 1815 novel *Guy Mannering*, Sir Walter Scott described it as "a mysterious, ill-defined consciousness that neither the scene nor the subject is entirely new." David Copperfield, in Charles Dickens's novel of the same name, experiences it not once but twice in the narrative. Around two-thirds of people experience ordinary déjà vu, which the psychologist Alan S. Brown has described as a "routine memory glitch" with various plausible explanations. One possibility is that we are experiencing a match between a genuinely new experience and an implicit memory for something we have experienced before, but for which we do not have explicit memory:

a dream, perhaps, or a familiar context for which we don't have an auto-noetic episodic memory. In line with this interpretation, it proves possible to elicit déjà vu experiences in the lab, by showing participants stimuli so briefly that they don't consciously perceive them, and then presenting them again for a longer period. One neurological explanation of déjà vu is that it is caused by brief random firings in areas of the brain that mediate the feeling of familiarity, such as parahippocampal and perirhinal areas in the right hemisphere. Patients with temporal lobe epilepsy, where such seizures are characteristic, frequently report déjà vu experiences during the "aura" that precedes the seizure. Another possibility is that transmission errors in the brain lead to a temporary increase in awareness that is interpreted as familiarity.

But Moulin's patients are not just experiencing pathological familiarity. Their feelings of previous experiencing are richer than the simple sense of having encountered a particular scenario before. Rather, they come with recollective experience, a sense of the self in the past. AKP actually *remembered* the events that seemed so familiar; he had a sense of himself living through them. That's why Moulin and colleagues prefer the term *déjà vécu* ("already lived") to the usual term *déjà vu* ("already seen"). When Moulin and his colleagues tried to confirm these informal reports with psychological tests on two patients (one of whom was AKP), they produced a large number of false positives on tasks of recognition for words and photographs. That is, they said they had seen items before when in fact they were being presented with them for the first time. They also reported high levels of recollective experience for items for which they falsely claimed recognition. They weren't just incorrectly rating them as familiar; they were actually saying that they remembered seeing the items before.

Moulin and colleagues interpreted their findings in terms of these patients having an overactive feeling of remembering. If you have persistent déjà vécu, you experience each new moment as if you were remembering it. Although AKP's diagnosis was uncertain (he died a few years ago), brain scans showed that he had abnormal levels of atrophy in his temporal lobes and hippocampus, particularly in the left hemisphere. Could it be that the "feeling of remembering" is centered in this part of the brain, and that in cases of persistent déjà vécu it is chronically overactive? This is the region, after all, that is affected by the seizures of temporal lobe epilepsy, which is also characterized by rogue feelings of re-experiencing.

In his most recent attempt to explain the neural mechanisms behind this memory abnormality, Moulin and his colleagues have proposed a new cognitive-neuroscientific model of déjà vécu. Central to their explanation is the existence of a particular cycle of activity in hippocampal neurons, known as the theta oscillation. (We have already seen how hippocampal theta appears to play a role in providing the "timekeeping" function in navigation.) Theta brain waves are relatively slow, high-amplitude oscillations in the firing of nerve cells, occurring at around six to ten times a second. Compared to other brain waves that you observe in an EEG, theta waves are deep ocean rollers rather than little ripples. It has been suggested that this distinctive theta oscillation allows cells in one portion of the hippocampus, known as CA1, to separate out their dual roles of encoding and retrieval. Like any wave, theta has peaks and troughs. It is the separation between these two phases of the wave that allows the hippocampus to switch rapidly between encoding and retrieval. Put simply, the model holds that a signal received at the peak of a theta wave is taken as relevant to encoding (the processing of new information), while signals that arrive in the trough of the theta wave are associated with retrieval. If there is a slippage in phase (such that peaks become read as troughs, and vice versa), arriving information that should be prepared for encoding may erroneously be read as relevant to retrieval. In other words, the representation of a new experience, which should be decoded as "This is new and happening now," is mistakenly read as "This has happened before."

The reaction to the publication of Moulin's déjà vécu results soon confirmed that this was not a one-off phenomenon. The power of Google recently led him to be approached by a woman in Dublin whose eighty-nine-year-old father, Patrick, had recently started exhibiting the symptoms of persistent déjà vécu. The fact that these experiences were particularly strongly associated with new experiences appeared to confirm the hippocampal theta hypothesis. In the topsy-turvy world of déjà vécu, when a new stimulus strongly activates the encoding system, it actually ends up strongly activating the retrieval one. For example, watching live TV with his daughter, Patrick would complain that he had seen the program before. His daughter described it like this: "We have quite funny conversations where I say, 'But, Dad, the guy on the TV is live, he's talking about *today*,' and my dad will say, with enviable logic, 'It was *today* when they filmed it. It was *live then*. He's saying *today* because it was *today* then! But I saw it yesterday.' "

Patrick is not an amnesiac like Claire; he has a good memory for his personal past, and shows the standard reminiscence effect, whereby events from one's late adolescence and early adulthood are recalled more strongly. The precise extent of his neuropsychological difficulties (if there are any) is still being investigated. But what interested me about Patrick's case, when I spoke to his daughter about it, was how he tries to rationalize his weird experiences. Confronted with a situation that he knows he cannot have encountered before but that nevertheless strikes him as familiar, he concocts a story to explain the anomaly. Television programs look familiar because the station is showing reruns to save money in hard times. "Those TV people," he has been heard to observe, "get away with murder, putting the same stuff on every day." Patrick has also become aware that others have noticed his behavior, and has become good at hiding it. A fanatical golf fan, he no longer watches coverage on the TV because he cannot enjoy it. He doesn't complain out loud about having seen all the games before; he just goes and does something else. To his family, Patrick's abandonment of his favorite sport is the clearest evidence that he is still experiencing déjà vécu. His story gives us a poignant insight into what it might be like to look back at the present from the standpoint of the future, because one cannot properly recognize the past.

WHEN MEMORY IS DISORDERED, CONFABULATIONS often follow. Moulin's patient AKP would explain how he had already read the morning paper (which had actually only just arrived) by going out in the middle of the night to read it in the shop. When his wife once found a coin in the street, he explained his feeling of déjà vécu by claiming that he had placed it there himself. Even his wife became a topic of his mnemonic storytelling. He stated that he had married her, the same woman, three times, in separate ceremonies in different parts of Europe. On a visit to the cinema to watch a film that he was (naturally) convinced that he had already seen, he explained the anomalous feeling by claiming that the film was actually about him.

In many ways, confabulation is a perfectly understandable response to these odd feelings of remembering. Simply imagining a possible explanation for a weird experience might, through a source monitoring error, lead to it taking the form of a genuine memory. If there is concomitant damage to the control systems (predominantly housed in the prefrontal cortex) that usually police the outputs of the medial temporal lobe system, the "memory" might become firmly entrenched. If the relevant monitoring and control processes in the prefrontal cortex do not reject that imagining as spurious, it can come to be experienced as having the truth of a memory, and trigger a cascade of story making.

A second theory of confabulations proposes a failure of strategic retrieval, particularly in the systems (also housed in the prefrontal cortex, but in this case more laterally) that organize searches of memory and monitor the outputs of those memory searches. Again, strategic retrieval deficits cannot be the whole story. Frontal pathology is not always evident in confabulation cases (AKP had no damage in that region, for example), suggesting that other neural systems must be involved. The most comprehensive theories of confabulation now propose "core deficits" in certain key functions: the intuitive feeling of rightness of the memory; the monitoring of the outputs of the memory "editor"; and related control mechanisms that decide whether the memory should be acted upon.

This multiprocess view of confabulations would explain why déjà vécu confabulations have their distinctive recollective quality. AKP did not only believe that he had married the same woman three times; he actually remembered having done so. Having come up with an explanation for his anomalous experience, the monitoring systems that would usually reject it as implausible failed to weed it out. His overactive feeling of remembering led him to experience his explanatory story as a genuine memory. In other cases of confabulation, the patient actually goes on to act on the basis of the wonky beliefs. Patrick, as we have seen, went so far as to avoid his favorite pastime, watching golf on the television. There was no point in switching it on, he would reason, because he already knew the outcome of all the games.

Martin Conway has noted that confabulations often have a particular emotional valence, frequently working to show the ego in a good light. As memory does more generally, confabulations serve the needs of the self. One of Conway's colleagues, Ekaterina Fotopoulou, has shown that the "self-serving" bias in confabulating patients is stronger than it is in the memory distortions of healthy volunteers. The weird stories generated by neuropsychological patients thus give us another illustration of the constant tussle between the forces of correspondence and coherence. Memory wants to be true to the way things are, but it also wants to tell a story that suits the teller.

FOR THE MOST PART, CLAIRE'S memory problems don't come with this kind of runaway storytelling. She does report some déjà experiences, but they have a rather unexpected focus. Claire experiences déjà vu (strictly, *déjà entendu*) for music, and not just for any music. She will hear some music that she cannot possibly have heard before, and she will think that it is a cover version, some old song being rehashed. And the songs in question have a particular quality. They are always "something by Elvis."

On the day I visit, Catherine Loveday is conducting a study with Claire on this aspect of her musical memory. She begins by reminding Claire of some previous studies they have done on music, including how she was asked to take notes on any memories stimulated by listening to old songs. As in the study of music-triggered involuntary memories that used popular songs taken from the preceding ten-year period, the songs that Claire hears on the car radio can trigger some general memories of her past, although specific episodic detail is quite rare. One exception comes from the songs of her second favorite artist, John Otway. Listening to one of his songs on an old LP, Claire wrote down how she could "visualize him singing and cracking his head on the mic, making real noise during his headbutts and all the atmosphere of everyone cheering." Interestingly, all the songs that she has noted as triggers for memory date from before her illness. She has not kept up with contemporary music at all, and so these tracks are potentially valuable entry points into her period of densest amnesia.

Even more intriguing, though, is this rogue sense of familiarity. Among her box files, Claire has found some notes written by her daughter, Georgia. Catherine reminds her that Georgia was present when they did their previous session on music, and Claire is pleasantly surprised to hear that her daughter has become involved. Georgia's notes pick up on the same phenomenon that has piqued Catherine's interest. "Claire hears a new Take That song," she writes, "however thinks it's eighties. Interesting. Just predicting predictable words or does she actually know the songs?" At one point Georgia's brother Leo passes through the kitchen and picks up on our conversation. "She recognizes everything," he tells us. "Every song. She says, 'This is a cover,' and we say, 'No, it's not!'" Catherine tells us that on other occasions when Claire feels that she doesn't quite know the song, she confabulates that it is just a bad cover of one of her favorites. "She says, I know this song but it's not being done prop- erly." Claire laughs at this. "It's like they're covering it," Catherine contin- ues, "and they're not covering it properly. So there's a sense of familiarity there, that it *feels* like it's a song that you know, but it doesn't sound right."

This is the wayward sense of familiarity that Catherine wants to explore in today's session. She has put together a CD with a selection of different kinds of song: some original Elvis songs, some Elvis covers with varying degrees of fidelity to the originals, and some original Beatles numbers. There are also some new numbers that Claire cannot possibly have heard before, as Catherine's musician husband has only just written them. These have been specially written to be "soundalikes," reminiscent of a particular style. Claire is told that she will be asked whether she rec- ognizes each song, and whether there are any specific or general memo- ries that come to mind when she hears them.

Claire's Elvis fixation is as obvious as I have been led to expect. For several of the Elvis songs she has an immediate, undisguisable reaction; she will start singing along, bobbing around on her kitchen bench with a childlike delight. "Not specific," she says, when asked for the memories triggered by "Jailhouse Rock," "but I would say definitely in my teens. Sheffield." Another confidently recognized Elvis track, "One Night with You," triggers something vague. "More recent feelings, but nothing spe- cific. It feels more . . . since my illness, memories of its warmth." "All Shook Up" sparks some fragile images of dancing with her husband, Ed. Some of the Elvis covers she quickly recognizes as having been the King's originally, while expressing disapproval at the modern interpretations. With the Pet Shop Boys' version of "Always on My Mind," for example, she is very clear that it is an Elvis song sung by someone else. The Fine Young Cannibals' take on "Caught in a Trap" has her pleading with us to stop the CD. "That's a *bad* copy of Elvis," she complains.

But some of the new songs, especially those soundalikes performed in a rock 'n' roll style, are also judged to be Elvis numbers. She will bob her head and sing along to them, even though she is only guessing at the words and cannot possibly have heard the songs before. After the clip has been played, she will say that she confidently recognizes it. "I know that it's an Elvis song," she says of one of the newly composed sound-alikes, "and I know that I know it very well and I must have it on one of my records." What about actual memories? "I can see myself putting my record on, my LP. Not just hearing it vaguely." For this track, she gives a score of four (out of five) to denote her certainty that it is Elvis. With another of the soundalikes, she tells us that it is similar to "Good Rockin' Tonight," but she doesn't feel sufficiently confident to attribute it to Elvis himself. A third soundalike is not explicitly recognized as an Elvis number, but she does have a feeling that she would find it on one of her many Elvis albums.

More bizarrely, Claire does the same thing with the Beatles numbers. She hears the famous opening of "Hey Jude" and confirms that it was originally performed by Elvis Presley. She isn't so sure in this case that the singer is Elvis, but she explains that that might be because she is used to hearing it on a record, while this recording is on a CD. "It's not the same, at all," she tells us. "It hasn't got the right depth and bass that you get from playing it on my record player. Which nobody else in the family, you know, really believes at all." "A Hard Day's Night" and "Help!" provoke the same reaction. In contrast, she is less than confident with at least one of the Elvis originals, and she fails completely to recognize the singer's voice on one of her favorite tracks, "All Shook Up."

When told of her errors, Claire is dumbfounded. She cannot understand how she could have falsely recognized so many songs, and she asks Catherine if she is allowed to hear them again. "It just feels wrong," she says. "I can't believe that I would have said I recognized a song. And did I think they were Elvis?" There is clearly a problem with her recognition of famous voices, but a greater issue is the rogue familiarity. She finds music familiar that cannot possibly be familiar. Some general memories are elicited by the songs, but there is not much of a specific nature. Since there is generally no recollective experience, her experiences cannot strictly be taken as examples of déjà vécu. One of the soundalikes, a sad ballad with a rejection theme, triggers a memory of listening to something similar while doing aqua aerobics, and being moved by the sentiment. But Claire knows that it is not the same song; it is simply triggering another memory through association, particularly of the emotions involved. When asked whether this particular song is an Elvis one, she is fairly clear that it is, and even quite confident that it is him singing it.

Catherine thinks that Claire is relying on the lyrics more than the music in making judgments about familiar songs. The soundalikes have fairly predictable words, so the fact that she can often guess what comes next might feed into her sense of familiarity. Catherine also notes Claire's immediate, tangible reaction to those songs that she finds truly familiar. "There is a very different level of reaction to the stuff that you absolutely know," she tells her, "compared to the things that you think you know." Later, Catherine tells me that she is currently following up this work to explore further the relation between musical misrecognition and rogue familiarity in Claire's musical experiences. She notes with amusement that she is in the enviable position, for a cognitive neuropsychologist, of planning an academic study based exclusively around the works of Elvis Presley.

At the end of my visit, I have the opportunity to watch Claire reviewing some of her recent SenseCam images. When she was first diagnosed, Claire was depressed and anxious. She was described as feeling overwhelmed and tearful in ordinary situations, frustrated at her inability to accomplish everyday tasks, and feeling a foreshortened sense of the future. As Demis Hassabis and others have shown, amnesia patients have difficulty imagining the future as well as recollecting the past, and so Claire's depression was likely exacerbated by a specific cognitive impairment. Neuropsychologists have described the distress caused by this kind of brain damage in many patients, as they become more aware of the mismatch between the self they used to be and the self they are now. If you cannot remember, you cannot update your self-image as your life moves on. You are caught between two identities, without the ability really to inhabit either.

The SenseCam gives Claire a way out of this impasse. I have been told that Claire will look at image after image without a flicker of recognition, and then something in an unremarkable picture will tip her over into remembering. Catherine is convinced that the sequential nature of the images is crucial; single images in isolation don't appear to have the same effect. "It feels," she tells me, "like the cues themselves are special because they are from her viewpoint and related to changes in her environment (movement or light), and therefore related to moments of particular attention, but also that bombarding the brain with these stimuli one after the other is key to her having these Proustian moments of recall."

I see this for myself on the day of my visit. One thing that strikes me immediately about Claire is how well organized she is: she has notebooks everywhere, including in the car. Her huge farmhouse kitchen table has a corner taken up with a workstation made up of notebooks and box files. "I have a pen and paper with me the whole time," she tells me, "and people accept that I'm scribbling, but it's not always easy in a group situation to make notes, and I'm always very upset that I don't remember enough. You like to remember people's, you know, things they say to you." Her work with the SenseCam (which she has been wearing around her neck since we arrived) is really just an extension of these scribblings. "You feel a bit self-conscious," she continues, "like, other people wouldn't react normally because they think that you're filming them." She keeps a separate diary to catalog her SenseCam sessions, so that she has some idea of the facts of the events she has recorded. She even draws little maps of seating arrangements, so that she can recall who was sitting where. Then, when she views the images and has those flashes of memory about the event, she knows what the autobiographical context is, those facts of personal history that frame all our memories.

This involves a fair amount of detective work. Today Claire is viewing some images from a week or so ago, when some friends came to stay.

Before she starts up the image viewer, Catherine asks her what memories she has of that visit. Claire remembers that there was some discussion about the family's two pet snakes, which belong to her now grown-up son and need a new home. But little else comes to mind. The first image in the viewer is of the very kitchen table we are sitting at, rather dimly lit in the evening. "I think we're eating pizza," she says, flicking through a few more images. "Making salad. Oh, no, I'm not. Looks like a cabbage I'm attacking." She scrolls through until she sees some more people. She wonders why they were having pizza. They would never normally have takeaway pizza except on a special occasion. Then she sees an image of someone holding, in some oven gloves, a fish-and-potato pie. "Oh no, that's right. I gave in and got a Domino's pizza in. I rang up for it and I went to get it for them." But why, when she had already cooked a fish pie? Then she remembers: they were celebrating. Georgia had got good GCSEs, and they were having a treat to congratulate her.

"Would you say that you are recalling this," I ask, "or are you working it out?"

"No, I wouldn't have recalled it. I'm working it out. But as soon as I saw the pizza box, I remembered."

This is one of the "Proustian moments" that Loveday and Conway have written about. There is a period of hard detective work, a sequence of logical inference, followed by the return of the genuine recollection: information that could not have been inferred, only recalled. Part of it, for Claire, was the emotion: the feeling of giving in to the request for a treat, after she had gone to the effort to make a pie. She jokes that she will soon even recall what toppings they had. "That must be Ed," she says, pointing at the figure of a dark-haired man in one of the images. The context helps to constrain his identity; in a public place, she would have much more difficulty recognizing him. In one image she sees herself walking over to the kitchen dresser, and recalls that she was looking for some Blu-Tack. That triggers a memory: the little girl from the visiting family was drawing a picture of one of the snakes, which was slithering around on the table at the time; if we turn around in our seats right now, we can see the child's picture stuck to the cupboard behind us. The different pieces of information are built up into a whole, a memory. "Like a jigsaw puzzle," Claire says. "What you're trying to do is put all the different bits together."

I ask whether SenseCam ever triggers any memories from before her illness. The answer depends on where she is. Although it wasn't directly related to SenseCam, she tells us about another Proustian experience she had recently, at an old-fashioned sweetshop in Covent Garden, one that had been furnished in a deliberately nostalgic way to remind people of the sweetshops of their youth. "I went in there and it was just like the whole . . . Although I couldn't smell it and I couldn't have tasted it, it was just the massive rush of warmth and memories." The kind of involuntary, sensation-cued memory that we all have for much more distant contexts is, Catherine observes, the same as what Claire experiences when she looks at SenseCam images from the previous week. "Yes," Claire agrees, "which is what I've lost, naturally. I could actually see that shop in my mind, and even the whole way down to it. It was such a big thing in my childhood."

It's impossible to know whether SenseCam is really helping to restore Claire's memory. My sense is that a lot of what she is doing is inference rather than real remembering (that *must* be Ed, rather than it *is* Ed). But I can also see that some genuine new memories are being unearthed. One reason for being skeptical about the effects of SenseCam is that it is impossible to know whether Claire has actually lost the memories she is trying to retrieve, or whether they have just become inaccessible as a result of the damage to her brain. Catherine thinks that the one-sidedness of Claire's lesion, confined as it is to the right hemisphere, points to it being more of an access problem than an encoding one. "When you can give the right cue, it can be surprising how much stuff has come out. And for Claire it's just a wonderful moment, to think that that stuff is not there, and then for it to come flooding back."

I also ask Claire about the future. When I think about catching a train back to London, I tell her, I can picture myself alighting at King's Cross, and imagine my feelings about what I have lined up for this evening—all those personal, episodic details that make the experience mine, even if only in anticipation. Claire is initially upset to find out that we came here by train. "I thought you must be driving," she said. "I would easily have come to get you." We gladly accept an offer of a lift to the station on the way back, especially since Claire is heading to the hospital for an appointment and she can drop us on the way. Can she picture herself driving into the hospital carpark, locking the car door, walking through the grounds and into the ward? "I can picture it well," she says, "because it will be the second time I'll have been." Something, though, stops me from being entirely convinced. Claire remembers some personal semantic facts about why she is going, and the instructions she has been given about parking, but that's not necessarily the same as saying that she can preview her future experience. I will not get a chance to ask her more about it now, but I would like to know if she is really placing herself in the future construction, starring in the imagining.

What about imagination more generally? How would she perform if she had to make up a story for a child, for example? I sense her starting to get upset. "I'd give it a go," she says. "I've tried that very sort of thing to amuse a small child, and not had confidence that I was doing it rightly for that age group, and knowing perfectly well that I could have done it much better." We are not sure whether this has to do with the amnesia or simply because, after a nursing career spent working with children and babies, she is not around kids much anymore. Catherine points out that their scanning studies with Claire have shown a heavily frontal pattern of activation when imagining the future, compared to what you would see in an intact brain. There is much less use of the classic medial temporal lobe memory areas that most of us would draw on in simulating future scenarios. It looks like a brain that is trying very hard, but doesn't have the raw materials it needs. The future events that Claire generates are also more generic and routine, and don't have the focus on emotions and possibilities that you would see in healthy participants.

The real benefits of SenseCam may be the personal ones. Memory is not just about remembering the past or predicting the future; it is also a way of being with other people. If you lose memory, you lose that opportunity to connect. As Claire put it to me earlier, an amnesia friendship is never more than one half of what it should be. Knowing that she is going to meet a particular acquaintance, she will often prepare for the meeting by reviewing SenseCam images of their previous encounters, so that she has a chance of sharing those common experiences. Even if she doesn't look back at the images very often, she knows that they are there. "It's the security of having it," she explains, "and knowing that I'm going to look at it all, big-time, one day." It gives her a greater confidence about moving through the physical and social worlds. Feeling happier and more socially connected, she can steer clear of further anxiety and depression and the memory distortions that they themselves bring. That makes things happier for her husband and children, and allows Claire to focus better on the various projects that keep her busy, such as volunteering for the Encephalitis Society and caring for some elderly neighbors.

Claire does drop us at the station, as she has promised. She remembers the route without any problem, although Catherine tells me that she tends to tell the same stories about the landmarks on the way, unaware that she has told them on previous occasions. She waves us off at the station with a warm, ever-so-slightly mournful smile. For the last four hours she has existed completely in the moment, remembering exactly who we are and why we are here; whether she will still retain any connection to it tomorrow or next week, we don't know. If I had not been told about her problems, and had not noticed all the notebooks and scraps of paper reminding her of the things she needed to do, I'm not sure I would have guessed that she was amnesic. For that, she can thank some extraordinary personal qualities, a humbling resilience, and the little bit of help she has had from the SenseCam. She is determined to put her life back together, and she is succeeding. "That's what those pictures do for me," she says, as the September breeze bangs at the car door. "They put me back together with the person I was before. It's hard work, but somehow they reconnect me to myself."

10

THE HORROR RETURNING

AS YOU BRANCH off the A1 motorway at the Sedgefield junction, the roar of one of the busiest roads in northeast England gives way to country lanes divided from farmland by hedgerows and grass verges. On a spring day in 2009, Colin was driving his thirty-two-ton tanker from the sewage treatment works at Darlington to the processing plant at Spennymoor. His job was to transport slurry to the larger works, where it was dried and compacted to make agricultural fertilizer. He would usually deliver five loads in a shift, and this was his third of the day. It was a Wednesday, around a quarter to ten in the morning. Most days, after his third load he would stop and have a break and a cup of tea with his mates at the works. Today he was deliberating on whether to stop earlier, at the snack van on the A1, but he decided to carry on past the van and not stop until he had made the delivery at Spennymoor.

He turned off the A1 and headed along the country lane that would bring him in a few minutes to the sewage works. As he drove, he noticed a car coming toward him from the north. It was a sky-blue Nissan Micra driven by a man in his early seventies, and it was drifting over onto his side of the road. As it came closer, Colin noticed the man's head nodding down and to the right, as though he was trying to find something lost in the passenger footwell. Colin flashed his lights and beeped his horn to try to get the man's attention, but he was still looking down at the floor of the car. Colin swerved to avoid the car, banking up onto a steep grass verge, and felt the eight-wheeler, with its full load of sludge in its tanker, starting to tip. The driver of the car was still not looking at him. Colin was as far up as he could go onto the verge, and there was nothing more he could do to avoid a collision. The Micra hit him head-on, directly under his driving position. It spun around through half a turn so that it was facing the same way as the tanker. Colin jumped down from his cab and went around to the other side of the crumpled car to see if he could help the driver. The windshield was out and the driver's window smashed. The man had been wearing a seat belt, and the airbag had triggered. Colin remembers seeing a scratch on the driver's nose where the airbag had impacted. "Can you get me out?" the man said to him. "Help me get out."

What happened next was a blur. Cars were stopped on both lanes of the road. There was a vile smell of fuel and hot radiators. Colin went back around the side of the lorry and threw up. The fire brigade came, and an ambulance. Colin was put into the ambulance and driven a short way up the road. They checked on him, made sure that he was unhurt, and left him while they went to attend to the injured man. Colin heard the sound of a helicopter, and a moment later saw an air ambulance landing in the field nearby. When one of the paramedics returned to the ambulance Colin was waiting in, he was able to ask whether they had got the old man out. Not yet, they said. It would take some time, as the driver was going to have to be cut free. It took forty minutes. In the meantime Colin was taken to the police station to be interviewed. As he was giving his statement in the interview room, a message came through on the officer's radio to say that the man had had two heart attacks in the ambulance and had died in the hospital.

The nightmares began straightaway. In his dreams, Colin saw the car weaving across the road toward him. He smelled the disgusting fuel smell, the exploded radiator. He got out of the lorry and saw the old man's face framed by the broken window. He saw the redness from the impact of the airbag. Rosy red cheeks, a bright red nose. He woke up shaking, in a sweat. He had been kicking his partner in his sleep, stamping on an imaginary brake pedal. Her leg in the morning would be black-and-blue. The accident had been his fault. He should have stopped at the snack van on the A1 and had his cup of tea. He had caused the crash, and the man's death. These thoughts kept him from eating. He would spend days at home, wandering around in a daze, mulling over what-ifs. He took up smoking again, and would be out there in the garden in the early hours, taking refuge from the nightmares, from the horror inside his own head.

Colin's employer was supportive, and he returned to light duties after a couple of weeks. One of his friends suggested that he get back in the wagon as soon as he could, and so he would sit in the passenger seat while one of his mates drove. He was coping. The inquest was scheduled for August, and he was asked to attend in order to give evidence. He had a holiday booked but he canceled it so that he could go to the hearing. He said to himself that he wanted to find out the facts about what had happened. He had not had any communication with the police since the accident, so he did not know whether he was being officially blamed for the crash or not. The court was told that the old man had been over the legal alcohol limit, and his family confirmed that he had had half a bottle of whisky the night before. No one—the police, the coroner, the deceased's family—blamed Colin for the accident. But he blamed himself. He should have stopped for that tea break. He should have taken a different route. It had been his first accident in fifteen years of professional driving.

His doctor prescribed mild sedatives, but they did not stop the nightmares. If he saw a sky-blue Nissan Micra, he would panic. Someone living farther along the street had one of these cars, and if Colin happened to glance out of the window as the car was going past, it would trigger flashbacks. He would feel nauseated, his palms would sweat and his heart would pound. He couldn't drive his own car, much less a lorry. He would press on the clutch but his leg would shake so much that he couldn't operate the pedal properly. He would sit outside the house for half an hour, tremoring spastically inside his own undrivable car, trying to put the thing into gear.

EMOTION DOES STRANGE THINGS TO memory. It is a basic fact of remembering that emotional events are remembered more clearly and in greater detail than neutral ones. They may also stick in our minds for longer. We can often recall events from childhood when they were humiliating or hurtful, but they are generally less accessible when they are emotionally neutral. There may be good evolutionary reasons for our being able to recall events that were threatening to the self in some way, especially if the real function of memory is as much to predict the future as it is to keep a faithful record of the past. When bad things happen to us, we learn from them, so that we don't make the same mistakes again.

One of the most thoroughly investigated examples of the emotional enhancement of memory concerns those snippets of our past that are bathed in light by the flash of historical events. I can recall in vivid detail the moment when, standing by a swimming pool in Spain on a September morning, I heard someone say that a plane had crashed into the World Trade Center. On a gray Sunday morning a few years before, I was in a supermarket browsing the newspapers when I read that Princess Diana had been killed in a car crash in Paris. Most of us will be able to recall occasions when shocking events seemed to leave an indelible mark on our memories. A psychological study from 1899 described distinctive flashbulb memories in people recalling how they had heard of the assassination of President Lincoln, more than thirty years earlier. The psychologists Roger Brown and James Kulik coined the phrase *flashbulb memories* in 1977 to describe our ability to recall not just the event of hearing the news but also our personal context at the time: where we were, what we were doing, who we were with. It is an unfortunate coining in one sense, as the authors admit, as it seems to support the mistaken analogy of memory to a camera. Rather, Brown and Kulik meant the phrase to capture the indiscriminate inclusion of contextual details. The flashbulb of memory illuminates everything in the vicinity, and it does so surprisingly and briefly.

Flashbulb memories lend themselves to scientific study because recollections can to some extent be matched to historical records of events. If we ask people how they heard the news of the death of Osama bin Laden, for example, we can corroborate certain aspects of their recollections with the facts of when and how the news was revealed. But information about personal context is much less easy to confirm. I happened to be watching the bin Laden events unfold live, in the early hours of a sleepless night, but anyone investigating my memory would have to take my word on that.

One way to confirm the unconfirmable is to ask whether people describe their flashbulb memories in the same way from one time point to the next. For experimental purposes, researchers have applied certain objective criteria in defining what should count as a flashbulb memory. One British study asked first-year university students what they remembered of the resignation of the prime minister Margaret Thatcher, within two weeks of the event. They were then reinterviewed eleven months later, and their memories were coded for details of how the news was learned, the people involved, the place, what the people were doing, and the source of the news. Nearly 90 percent of the participants recalled the event with very high consistency at the retest: that is, showing minor inconsistencies in no more than one of the five categories.

This strong empirical evidence for the existence of flashbulb memories has led some to suggest that they represent the operation of a special memory mechanism. Brown and Kulik drew on an earlier neurobiological theory, according to which the brain, stimulated by the exciting event, sends a "Print now!" command to encode everything into memory. But there is no strong evidence that such a mechanism is at work. Flashbulb memories are just as susceptible to distortion and storytelling as ordinary memories. When 106 university students were asked about the 1986 *Challenger* space shuttle disaster a few hours after the event, they could clearly remember how they had heard the news. Two and a half years later, their "flashbulb memories" showed plentiful evidence of forgetting and distortion. A quarter of the students reinterviewed were wrong about the main facts of their original reports. Whereas only nine of the undergraduates had initially reported having seen the news on TV, nineteen of them claimed as much at the later date. At the same time, many of the participants were confident about their memories, and were surprised when they were told, in a later interview, about the extent of their memory distortions. It may be that we are particularly confident about our memories for historical events because we intuitively believe some version of the "Print now!" theory. Alternatively, we may be especially certain about our memories because we are making errors in source memory, thinking we got the information from one source where actually we gleaned it from somewhere else.

The emotionality of the flashbulb event is bound to contribute to that runaway confidence. Because the happenings mattered so much, we feel more attached to any memories we have of them. Soon after the terrorist attacks of 9/11, a group of memory researchers from several different institutions got together to study how flashbulb memories of the atrocities persisted over time. The findings were similar to those of the *Challenger* study. After one year, only 63 percent of 9/11 memories were consistent with original reports (the number dropped slightly further, to 57 percent, three years after the attacks). But respondents' confidence in their memories remained very high. One of the researchers on the project, Elizabeth Phelps, put it like this: "Usually, when a memory has highly vivid details and you're confident in those details, that means you're likely to be right. Confidence often goes hand in hand with accuracy. But when something is highly emotional, they often get separated." We feel our reconstructions of the event with such emotional force that we cannot help but believe in them.

It seems likely, then, that the ordinary mechanisms of memory can account for the flashbulb effect. Flashbulb events are by definition surprising and distinctive, both factors that we know to increase the memorability of information. Flashbulb memories are often also highly rehearsed and talked about with others, which increases the likelihood that they will be retained. Above all, flashbulb memories occur for events that matter to us in some way. They are emotionally arousing, especially when they are relevant to the self. In the 9/11 study, the likelihood of forming flashbulb memories increased with proximity to Ground Zero: those physically closest to the events were more likely to have vivid memories for them. Our memory for events from our personal family history, such as births and deaths, can also show the flashbulb effect. Indeed, one study of American students showed that only 3 percent of flashbulb memories were about events of national importance; most focused on self-relevant events such as injuries or accidents, memories from college freshmen's week, and romantic encounters.

We know much more about the role that emotion plays in memory than we did at the time of Brown and Kulik's observations in the 1970s. When we are emotionally aroused, the amygdala is busy processing the emotional significance of the events we are witnessing. This activation leads to the release of hormones and neurotransmitters such as adrenaline, noradrenaline and cortisol, which form part of the body's response to stress. The presence of these substances in the hippocampus affects the synthesis of the proteins responsible for the consolidation of memories. In brain-scanning studies, researchers have found the amygdala and hippocampus working together and activating at the same time when emotionally charged material is being learned.

It is clear that the bodily systems that process emotion have a big part to play in modulating what we remember. But what happens when the emotional repercussions of an event are extreme and overwhelming, as in Colin's case? It has often been suggested that psychological traumas are remembered in a special way, through mechanisms that are not involved in ordinary remembering. Is that true, or is memory for trauma essentially the same as any other kind of remembering? We have seen that flashbulb memories, once thought to reflect the operation of a special memory system, are actually explicable in terms of basic memory processes. Likewise, a noisy debate in the science of memory has asked whether memory for trauma is really as special as has sometimes been thought.

THIRTY YEARS ON, PETER STILL remembers the Welsh padre. Troops from 45 Commando Royal Marines had been among the first to set foot on the Falkland Islands after the Argentine invasion in April 1982. Resting at a place called Estancia House after forty miles' yomp across the East Falkland moors, Peter and his comrades heard on the comms radio that four of their number had been killed by enemy fire. One was Peter's good friend Mike. As they were standing there absorbing the news, Peter saw the sergeant-major walking over to their small group and asking if they would come with him. The padre was alongside him. He was a colorful character, with a broad Welsh accent, very much one of the boys. Peter was told that he and his comrades were to form the burial party. They carried spades out onto the moor and dug a shallow trench. As they had no body bags, the four dead bodies were laid out in their hooded sleeping bags. The men removed their own berets and stood at attention while the padre conducted the funeral service. Then, with the dead men laid out side by side in the trench, the burial party began to fill the grave with soil. As Peter was filling the earth back in over the bodies, the hood of Mike's sleeping bag fell open with the weight of the soil, and Peter saw his friend's face. He had to put his spade down and walk away. He was in tears. The image of that moment stayed with him more vividly than any other event from the war. It was a memory that was to stay with him for three decades.

"War not only kills and wounds," observed a recent article on military trauma in the British armed forces, "it also generates some of the most intense stressors known to man." In the early 1980s, when the battle for the Falklands was fought, posttraumatic stress disorder (PTSD) was still a relatively new diagnosis. Originating in observations of "shell shock" and "battle fatigue" dating back to the nineteenth century, it was eventually formally recognized as a psychiatric disorder with the publication of the third edition of the psychiatrist's bible, *The Diagnostic and Statistical Manual of Mental Disorders* (*DSM*), in 1980. Although it is estimated that about three-quarters of people in the United States experience a traumatic event at some point in their lives, the lifetime incidence of the disorder is much lower, at around 8 percent. Plenty of people experience a trauma without going on to develop PTSD.

Despite this formal recognition, this diagnostic move by *DSM* has not been without its critics. Some, such as the psychologist Richard McNally, have pointed out that "trauma" is extremely hard to pin down. Although it is officially defined in *DSM* as provoking fear, helplessness or horror in response to a threat to one's life, the vagaries of subjective interpretation mean that individuals differ wildly in how traumatic they find the same happenings. An event that will scar one person for life will be shrugged off and forgotten by another. Conversely, PTSD diagnoses are occasionally made in response to events such as minor car accidents and overhearing sexual jokes at work, which many would judge unpleasant but hardly the stuff of trauma. Critics of PTSD complain that it is a social construction, a disorder invented by a culture obsessed with trauma as a key to human identity.

What is undeniable is that PTSD is at root a disorder of memory. The diagnosis is confirmed when vivid, uncontrollable memories of a traumatic event are so frequent, persistent and debilitating that they disrupt behavior and lead sufferers to try to avoid situations that will trigger them. People suffering from PTSD flashbacks and nightmares typically feel the returning experience with much of the emotional force of the original event, including physiological correlates such as sweating and heightened heart rate. That emotional reaction makes it difficult for the patient to become aware that the trauma lies in the past, and that it is not still happening. In the words of PTSD expert Rachel Yehuda, the combined effect of these symptoms is to make the sufferer feel "haunted by the past."

It is certainly true that the uncontrollability of memories sustains the disorder. The intrusions can occur at any time, even in sleep, and so it is almost impossible to defend against them. But people do try. They may go out of their way to avoid meeting people who remind them of the trauma, or to steer clear of places and events that commemorate it in some way. Peter, for example, would not fly in an airplane, and would stay away from Bonfire Night celebrations with their noisy fireworks. Sufferers may try to suppress their intrusive thoughts, which in a cruel irony is known to increase the likelihood of the undesired thoughts occurring. These active attempts to avoid thinking about the trauma can sometimes be successful, but at night, when the sufferer's cognitive defenses are down, the intrusions commonly resurface as nightmares. In the years after his return from the Falklands, Peter would often be woken by the vision of his dead friend's face emerging from the sleeping bag. Fearing that he too would be buried in his sleep, as he nightmarishly imagined that his friend had been, he would try to force himself not to lose consciousness again. In ordinary people, nightmares are mostly confined to REM sleep, the phase of sleep when the brain switches off connections to the muscles of the body, leaving us effectively paralyzed. In PTSD sufferers, nightmares can also occur in non-REM sleep, resulting in violent thrashing and (as in Colin's case) accidental injury to partners.

These distinctive features of PTSD have led psychologists to investigate whether it represents the functioning of a different kind of memory. The main focus of these efforts has been on a special form of memory intrusion known as flashbacks, where the sufferer re-experiences the event with such vivid sensory force, and with so many of the accompanying bodily sensations, that they feel as though they are reliving it. Flashbacks typically involve intense visual imagery, thematically congruent with nightmares, and last for about a minute. Richard McNally has related the story of a veteran who was suddenly propelled back to the horror of war by the sound of children's firecrackers exploding under the wheels of his jeep. "Although on one level he realized that he was in Colorado, not Vietnam, the emotional and behavioral reaction triggered by the firecrackers was the same he had exhibited years before during ambushes while in Vietnam."

And yet research has questioned whether these experiences are specific to traumatic events. In one study, the Danish psychologist Dorthe Berntsen asked twelve college students with a diagnosis of PTSD to record their involuntary memories. She then analyzed the first fifty memories in each case. Only just over 5 percent of them related to the trauma, and less than 2 percent qualified as true flashbacks. In contrast, more than half of the involuntary memories related to positive or neutral events. A tenth of the memories were classed as nontraumatic flashbacks, and were often intensely positive. Even in people with a PTSD diagnosis, flashbacks are not specific to trauma.

In a later study, Berntsen and her colleague David Rubin conducted a telephone survey with a large sample of Danes aged between eighteen and ninety-six. Their focus was on involuntary memories: those that popped into consciousness without warning. Positive involuntary memories outnumbered negative ones by a ratio of about two to one, similar to the proportions found in ordinary autobiographical memories. With increasing age, people reported lower frequencies of recurrent memories and dreams, but those memories became more positive and intense, also fitting with previous findings that negative emotions are experienced less frequently and with lower intensity in old age. Positive recurrent memories in respondents over forty tended to cluster around late childhood and adolescence, much as ordinary autobiographical memories generally do.

The researchers then asked about specific trauma memories in a subsample of the PTSD participants from Berntsen's earlier study. The volunteers kept a diary of their involuntary memories and were asked to code each one for its relevance to the trauma they had previously described. The average time that had elapsed since the trauma was just over two years. The memories were rated by independent judges according to whether they accurately replicated reports of the same event given earlier in the study. Although the participants clearly remembered the same event, they remembered different aspects of it, or different "time slices" of the narrative. Only one participant reported memories that were almost perfect reproductions of her earlier memories. In her case, though, the three reported memories all entered her consciousness when she was in an identical situation (jogging alone along an isolated nature trail) to that in which the trauma occurred, suggesting that the memories that came to her were heavily constrained by the cues available.

The inconsistent "flashbacks" described by Berntsen and Rubin are probably not the only way in which intrusive memories of trauma are subject to the same kind of distortions as ordinary memories. After careful weighing of the ethical pros and cons of their apparently extreme procedures (particularly the scientific value of learning more about these debilitating experiences), researchers have sometimes used harmless drugs such as sodium lactate to set off panic attacks, which have been shown in turn to trigger flashbacks. One such study of Vietnam veterans showed that the intrusive recollections that were triggered did not appear to match real trauma events. One veteran's drug-induced flashback concerned killing a Vietnamese woman who then rose from the dead. The intrusive image was clearly related to the war, but it concerned an event that could not actually have happened. In Peter's intrusive memory, his friend was sometimes still alive and pleading with him, "I'm not dead, don't bury me." Other studies have shown that people can have flashbacks about the murder of a loved one, even when they were not present when the horrible crime happened and thus did not actually witness the event.

Trauma flashbacks therefore present us with a paradox. They can be so strong and yet so wrong. One of the tenets of the reconstructive view of memory is that we remember events through a filter of our later emotional states, such that if we start to feel differently about a situation, our memories change. We also appear to construct flashbacks about what we feared *might* have happened rather than what actually did, such as in the image of the resurrected Vietnamese villager. Imagination inflation will thus also play its part. In one case study from the recent clinical psychoanalytic literature, a baffled gynecologist had to insist that a patient's clitoris had not been removed, despite her traumatic memory of its having happened. If we can believe that we proposed marriage to a Pepsi machine on the basis of having simply imagined it, how much more likely is it that our imagined fears can become fixed in our minds as "memories"?

So far, the popular conception about how memory works in PTSD seems to have little rooting in science. Richard McNally has pointed out that the term *flashback* actually originated in the movie industry, where it describes a device for articulating the components of a complex narrative. Another standard trope of movies and novels is that of the victim or perpetrator of a crime who cannot remember what happened. There is certainly plenty of evidence that trauma can lead certain aspects of the event not to be encoded properly. In the phenomenon of "weapon focusing," for example, victims of gun crimes can often give very accurate descriptions of the weapon while showing amnesia for other aspects of the context, such as the color of the attacker's eyes. But this is not remotely surprising. When someone is pointing a gun at you, you look at the gun; you don't pay attention to the decor. Selective memory for traumatic events is simply a normal result of the emotionality of a highly abnormal situation.

Others have suggested that PTSD results in more general failings of memory. People with PTSD can be forgetful, which fits with the fact that their general psychological distress hinders them from encoding new information. When asked about their memories of their earlier lives, their responses can be excessively generalized and lacking in specificity, a pattern also found in depressed people. Similarly, PTSD sufferers can have difficulty in richly imagining their future lives, which fits with the growing evidence that remembering the past relies on similar mechanisms to imagining the future.

Researchers have investigated the possibility that the memory difficulties associated with PTSD indicate the operation of a distinct memory mechanism that leads trauma to suppress other memories. Sigmund Freud proposed that forces in the unconscious can act to "repress" the memory of a trauma and keep it out of consciousness. In recent years, psychologists and neuroscientists have made progress in identifying how this kind of "motivated forgetting" may declutter our minds to allow more efficient thinking and remembering. In one study, researchers asked undergraduate students either to think about or try to suppress words that they had previously learned to pair with other words. The more the volunteers tried to suppress the associated word, the less well they remembered it. In brain imaging studies, memory suppression has been shown to be associated with increased activity in the prefrontal cortex (fitting with the idea that suppression is a process of active, effortful inhibition) and diminished activity in the hippocampus, consistent with the information being less well remembered.

Although evidence is building that memories can in some circumstances be successfully suppressed, there is no evidence that these difficulties are specific to trauma as opposed to any other emotional stimulus. The extreme emotion of the trauma may skew remembering toward specific features of that event, as in weapon focusing, but so do positive emotional stimuli. For example, one study reported that showing subjects a picture of a naked person in the middle of a sequence of clothed individuals disrupted memories for the less memorable images that followed. For most people, seeing a picture of a nude is not a frightening event, although it may trigger other feelings. In this experiment, these distinctive emotional stimuli seemed to impair encoding of the material that formed the background to the nude.

The balance of evidence suggests that trauma survivors do not differ from healthy people on their memory abilities. If there are gaps in the person's memory for the traumatic event, they may be explicable simply in terms of the effects on attention (and therefore encoding) of an extremely frightening stimulus. Any general forgetfulness subsequent to the trauma may be explained in terms of the distress of the disorder. Although PTSD sufferers do seem to show neuroanatomical differences, such as smaller hippocampi, it is not entirely clear that these are caused by the trauma. In a study of identical twins in which only one of the pair had been exposed to a combat trauma, the size of the hippocampus in the nonexposed twin predicted how likely the exposed individual was to succumb to PTSD. These findings suggest that hippocampal volume may represent one of the factors that predispose certain people to developing PTSD after a horrific experience, rather than being a result of that exposure.

So far, then, there is no evidence that memory for trauma operates in special ways. One of the most controversial versions of this view proposes that trauma can be "remembered" in ways that leave no explicit, consciously accessible trace. Some have proposed, for example, that traumas can be forgotten at the explicit level but retain their effects implicitly. Joseph LeDoux of New York University gives the example of the memory of surviving a car crash in which the car horn becomes jammed on. When the sound of a horn is heard again, representations in two memory systems are activated. The sound of the horn is implicitly "remembered" by the amygdala memory system and acts as a conditioned stimulus, triggering a bodily fear response. The neural representation of the noise also sets off the explicit memory system of the medial temporal lobe, and episodic memories of the event are activated. With both memory systems working in parallel, the experience of remembering the accident has both explicit and emotional qualities. You remember the car crash, and you also distinctly remember how it felt.

Sometimes, though, a cue can retain its power to trigger the amygdala system while no longer being effective as a cue to explicit memory. With the passing of time, for example, you might forget the explicit memory detail that the car horn stayed jammed on. In this scenario, you later hear a car horn and experience an unpleasant, mysterious fear response. You feel afraid without knowing why you are afraid. The horn has been "remembered" by your amygdala system but "forgotten" by your explicit memory. The trauma is not actually forgotten; it is just that certain cues have lost their power to trigger the explicit memory.

This is critically different from the idea that traumatic events can be banished from consciousness by subconscious forces. In the Freudian model of the mind, the forgetting of trauma involves the active efforts of forces in the unconscious to "repress" the memory and keep it out of consciousness, where it could be damaging to the ego. But there is no good scientific evidence that these unconscious forces exist. Traumas are remembered, and they are remembered only too painfully. They may not be thought about for a long time, perhaps partly as a result of conscious, successful efforts to suppress memories about the event or to avoid situations that might involve triggering cues, but they are not forgotten. You may forget that the car horn stayed on, but you don't forget the car crash.

The finding that pure emotional cues can trigger explicit memories is also not at all at odds with the modern scientific view of memory. As we have seen, PTSD sufferers can have their flashbacks triggered by bodily changes resulting from drugs like sodium lactate, as well as by external sensory cues such as the flashing lights of an emergency vehicle. But those veterans have never forgotten their trauma; they are just having their memories cued in different ways. A relevant clinical observation is that survivors of childhood sexual abuse can in some cases have their horrible memories reawakened by an event that puts them in the same extreme emotional state that they experienced during the trauma. A woman who has been date-raped under the influence of a drug like Rohypnol might have traumatic memories cued by the numbness that follows a surgical anesthetic. On a much milder level, a friend told me that being depressed now as a fortysomething reminds him of his often miserable childhood. Nothing about the content of his thoughts pushes him into making the connection; it is simply that the emotion recalls the emotion. Sights and sounds can serve as cues to memory, but so can feelings.

Any experience that leads to implicit remembering should also leave an explicit trace. On the basis of current scientific evidence, no special mechanism needs to be postulated to explain why memories for trauma are sometimes incomplete, or why emotional cues can be particularly powerful. As we have seen, this became much more than an academic debate with the storm over cases of "recovered memory," in which people engaged in certain kinds of therapy claim to be suddenly re-experiencing memories of horrific abuse suffered in childhood. In LeDoux's model, implicit memories for a traumatic event can persist, but they cannot be converted back into explicit memories unless those memories are already there. Few memory scientists believe that it is possible to suffer a trauma that is *only* remembered at the implicit level, until such time as the explicit memories can miraculously be unearthed. If a reasonable amount of cueing does not bring the memory to light, then there is no memory to retrieve. Rather, recovered memories of abuse (as opposed to those that return to consciousness spontaneously, through a normal process of remembering) appear to be fanciful constructions that can usually be linked to the suggestions of overzealous "recovered memory" therapists, with their emphasis on hypnosis, repetition and imaginative reconstruction—all techniques that are well known to lead to imagination inflation and other reconstructive errors.

There is no question that memories of trauma, like any memories, can be partial and incomplete. As we have seen with weapon focusing and other examples of memory narrowing, the emotional intensity of a traumatic situation can lead victims to focus on some aspects of the event at the expense of others. In the death camps of the Second World War, paying too much attention to the brutality around you was a capital offense. In Douwe Draaisma's words, "The injunction not to be conspicuous under any circumstances was followed by a second injunction: *Thou shalt not look*. A prisoner who was seen watching an SS man ill-treating a fellow prisoner was flirting with death."

As a result, some survivors complained of a kind of amnesia for these details, compounded by the effects of extreme malnutrition, violence and hopelessness. In the psychoanalyst and concentration camp survivor Bruno Bettelheim's case, this included "the ever present sense of 'what's the use, you'll never leave the camp alive'.... So both powers, those of observation and of reaction, had to be blocked out voluntarily as an act of preservation." This is not evidence for traumatic repression: it simply illustrates the truism that if certain details are not attended to and encoded, they cannot be remembered later. In every other respect, people remember the concentration camps only too well. The memory of Holocaust survivors is as prone to reconstructive errors as anyone's, but one thing that cannot be said is that people forgot their trauma.

The Freudian concept of repression has come under sustained attack in recent years, perhaps due in part to a lack of clarity in Freud's own writings about the extent to which repression mechanisms are actually unconscious. Claiming that repression only works for repeated traumas, as some therapists have done, is inconsistent with the overwhelming evidence that repetition enhances memory rather than diminishes it. If anything, repeated traumas will lead to strong *general* memories of the events in question that may nevertheless be lacking in specific detail. Daniel Schacter points out that he has considerable difficulty remembering the details of the many different airplane trips he has made. But he has excellent memory for the general event of flying, and he would never for a moment forget that he had ever flown in an airplane.

With traumatic repression largely discredited, many clinicians have turned their attention to the more strongly supported clinical concept of *dissociation*, according to which trauma causes the mind to become splintered into separate zones with separate memory systems. Survivors of rape, for example, often report forcing themselves to become distant from the events as they happened, which might conceivably lead to fractionated memory. The role of dissociation in memory disorders has been most controversial in the syndrome of dissociative identity disorder (previously known as multiple personality disorder), in which the patient creates multiple "alters," each with its own separate system of memory. Genuine cases of dissociative identity disorder are rare, and when they are thoroughly documented they count as a serious psychiatric disorder, albeit one whose cause is currently unknown. There is evidence that at least some cases of dissociative identity disorder should be classed as elaborate delusions. Some studies have shown, for example, that patients can remember information across the boundaries of their alters, suggesting that their "amnesia" is at some level feigned. Whatever the truth of dissociative identity disorder, the issue is not whether dissociation happens—it almost certainly does—but whether it constitutes a special mechanism for trauma memory.

We are a long way from a clear understanding of how memory works in cases of trauma. Some rare cases of psychogenic amnesia, such as those "fugue" states where a trauma can apparently lead someone to lose their memory (and thus identity) entirely for a short period, currently have no explanation in science. Saying that Freud's theory of repression is mistaken is not the same as saying that trauma cannot be forgotten, or not thought about for a long time. It is rather the issue of whether there is a dynamic subconscious force actively pushing out the memories. Such a force would contradict at least two basic laws of memory: the idea that repetition enhances retention; and the idea that emotional events are remembered better. There seems to be no good evidence for postulating such a mechanism, in which case we should be guided by the scientist's predilection for the simplest possible explanation.

In other respects, things will never be entirely clear cut. Much of the movement toward a reconstructive view of memory has been about rejecting the idea that memories are "possessions" that you either do or do not have. If memories are constructions, then they can presumably collapse back into their constituent elements, to be reassembled in different ways on other occasions. During the period in which the memory is not regularly coming to mind, it cannot be said to be forgotten. But, like my memory of the first fish I caught, it cannot exactly be said to be remembered either. It turns out that a crucial factor in suspected cases of recovered memory is whether the victim remembers remembering. If you say that you are experiencing a memory of an event for the first time, you are making a strong claim that you *forgot* that event in the intervening period. We are rarely so accurate in evaluating our past acts of remembering. My experience is of remembering something for the first time, but how can I know that for sure? My judgment is based on my ability to remember other occasions on which I remembered. I may in fact have recalled this event before, but forgotten having done so.

Jonathan Schooler of the University of California, Santa Barbara, has argued that this is an essential factor in suspected cases of recovered memory. He and his colleagues have used very strict criteria for what should count as a recovered memory: evidence that the traumatic event happened, evidence that it was forgotten, and evidence that it was subsequently remembered. In several of the cases he has studied, people who report a returning memory of trauma fail to remember that they mentioned the event to someone else during the intervening period. One case involved a forty-year-old woman who recovered a memory of being raped while hitchhiking. She reported to her therapist that the memory had only just reemerged, but it later transpired that she had in fact mentioned the trauma to her ex-husband on several occasions. There was no question about the accuracy of her memory for the assault having happened, but she seemed to have forgotten the fact that she had previously remembered it. What seemed to have changed in the meantime was the interpretation of the event: the fact that this unpleasant sexual experience had actually been rape. When we start to feel differently about an event, we also start to remember it differently. Schooler and his colleagues have dubbed this the "forgot-it-all-along" effect. We are demonstrably bad judges of our past abilities to remember things. Many "recovered memories" may actually have been remembered all along, without the subject recalling having done so.

What has happened to the memory during the "forgot-it-all-along" period? Some, such as Richard McNally and Elizabeth Loftus, would see the simplest explanation as being that it is possible not to think about something for a long time without actually forgetting it. If an event *has ever been* remembered, they would hold, then it cannot really be said to have been forgotten. This can seem to hold too much to an all-or-nothing, "possession" view of memory. If we instead follow the logic of the reconstructive view, we do not "have" memories; we construct them when we need them, on the basis of different kinds of information. It is therefore presumably possible for the elements of the memory to remain in storage while the construction—the full-blown episodic memory—is itself some way distant from consciousness. In that case, the usual process of cueing should ordinarily be sufficient to cause the event to be remembered. What is special about supposed cases of repression is that the memories are so inaccessible to consciousness that nothing brings them back, at least until the special conditions of psychotherapy are in place.

There are good reasons to believe that memories of trauma often exist in this fragmentary form. Arthur Shimamura of the University of California at Berkeley has observed how trauma can leave "free-floating" fragments of memory, only tenuously connected with contextual details of time and place, but with strong associations with emotional reactions. "With trauma memory," the clinical psychologist Kevin Meares told me, "what you've got are fragments and lack of coherence, and so things are parceled off, separated, and remain as unlinked, unrelated and frozen in time." The task of the trauma victim, ideally with help from an open-minded clinician or therapist, is to do what (in Daniel Schacter's phrase) all rememberers normally do: "knit together the relevant fragments and feelings into a coherent narrative or story." But those stories are unlikely to be any more accurate than ordinary memories, and there are good reasons for thinking that they will be less so.

Memory in trauma is slippery because memory in everyday life is slippery. Could the situation be any different for traumas that occur in childhood, when the autobiographical memory system is particularly fragile? We know that a traumatic event cannot activate the emotional memory system of the amygdala and related structures without also activating the explicit memory system of the medial temporal lobe. But what about when the explicit memory system has not yet developed, and the child is still in the grip of childhood amnesia? One theory proposed that it is possible in childhood to have implicit emotional memory without explicit memory because the hippocampal system is not yet mature. This is held to explain, for example, how childhood fears, stored in neural systems outside the hippocampus, can be suddenly reawakened by stresses in adulthood. Such an argument would seem to be contradicted, however, by the findings that children can remember much further back into their childhoods than adults can. If brain maturation were the only factor in our amnesia for explicit memories in adulthood, then asking children at different ages should make no difference.

Some have proposed that, without a fully functioning episodic memory system, the opportunity for "reliving" a memory will be lacking, and so memories of trauma might be manifested in other ways, such as through reenactment of the event in play. The therapist Lenore Terr has been particularly associated with the idea that traumatized children display their retention of the traumatic event through "behavioral memories." For example, a child who has been sexually abused might engage in inappropriate sexual behavior with dolls, or a kidnap victim might act out abduction fantasies with her teddy bears. The difficulty is in inferring abuse from such behavior in the absence of any independent corroborating evidence. There is currently little good scientific evidence that children "enact" memories in play that they cannot explicitly recall.

When it comes to explicit recall in childhood, memories for traumatic events seem to be forgotten and recalled in similar ways to memories for happier ones. Some children, such as the little cranial surgery patient Michael, appear to have good explicit memory for their trauma, and it seems likely that the stressful context of traumatic events enhances memory in children as it does in adults. But such memories do not persist into later life. In one study of children who had suffered a documented trauma such as sexual abuse or kidnapping, traumas that had happened before the age of around three were only sketchily recalled, if at all. The veil of childhood amnesia is drawn over traumatic events as much as it is over nontraumatic ones. Even with older children, there is little evidence that traumatic events are remembered in a fundamentally different way than everyday ones. Those that are recalled are often recalled for longer, but that may just be because traumatic events are distinctively emotionally salient, and distinctive events, good and bad, stick in one's mind.

Confirming this view, it seems that children's explicit memories of trauma are as prone to distortion as those of adults. Lenore Terr asked twenty-six child victims of kidnapping about their experience several years after the trauma. Eight of the children recalled the event quite accurately at the time, but all but one of these demonstrated memory distortions when interviewed four to five years later. For example, one survivor remembered there being a pair of female kidnappers, when in fact the culprits were all male. Far from being indelibly inscribed onto memory, traumas suffered in childhood are as prone to distortions in the retelling as any other memories.

I MEET COLIN ON A gray day at his house on an estate in a former mining village in the north of England. He is in his early thirties, with close-cropped hair and glasses, his slender but well-gymned frame bearing the slightly uncomfortable-looking color of a recent tan. Before his treatment, he tells me, he would never have dreamed of going on a foreign holiday. He was constantly in the house. He thought that if he went out of the house he would either hurt somebody or be hurt by someone. He couldn't go in the car to the shop, not even as a passenger. He was constantly mulling over what might have happened if he had made different choices: if he had stopped for that cup of tea on the A1; if he had taken a different route; or if he had simply not turned up for work that day. Home was the only place where he felt safe—from the certainty of more bad things happening, if not from the thoughts of them.

We sit and drink tea in his front room while my digital voice recorder encodes our conversation. The house is scattered with the toys of his young son, born a month after the tragedy. He begins by telling me the story of the accident: the blue Micra weaving across the road toward him, the smell of hot radiator fluid, the dying man's face in the car window. He is careful to separate out how he felt in the months following the accident—the nightmares, flashbacks and constant anxiety—from how he feels now, after his treatment. He is better, but not completely so. He sounds hunted and tense, still in obvious psychological pain. I get an uneasy sense that he feels that I am assessing him in some way, testing whether he can match up to some imaginary standard. But he has been keen to talk to me, to tell his story in the hope that a better understanding of memory for trauma can help others in the same horrible situation. He blamed himself for something that he should not have shouldered the blame for. When that changed, so did everything else.

After talking to a series of therapists and psychiatrists, Colin was eventually put in touch with Sitha, a psychiatrist who specializes in road traffic traumas. She replaced the mild sedatives he had been taking with antidepressants, and explained that she was going to start him on a course of therapy using a relatively new and yet seemingly effective technique known as EMDR. This forbidding acronym stands for *eye movement desensitization and reprocessing*, and it involves the patient following a moving LED display with his eyes, and tracking the lights' side-to-side movement as it is mirrored by the noise of handheld buzzers. Sometimes the weird-looking technical apparatus is done away with, and the client simply follows the horizontal movement of the therapist's finger. In a way it is a version of the old-fashioned hypnotist's swinging gold watch, and it seems to have some of the same psychological power.

A typical EMDR session lasts about half an hour. In their first full session, Sitha asked Colin to watch the lights and think about the sky-blue Micra, the object of his terrible anxiety. "Not every blue Nissan car is going to have an accident," she reassured him. Then, after a while, she asked him what feelings he had about these cars. "She was doing things," Colin tells me, "like step by step just trying to tease the information out of my brain, what I'd locked away in the back of my mind. Just to try and help pull distant memories out from the back of my mind."

When I went to talk to Sitha about how EMDR works, she told me that the process of following the light across both halves of the visual field somehow allows memories to become freed up, and for previously inaccessible details to be remembered. The memory becomes more fluid, more plastic, and can thus become more fully integrated. If memories are reconstructions, the claim goes, EMDR allows the rememberer to rebuild them as more complete representations.

At first, Colin was skeptical. "A little box like that," he says, remembering the first time he saw the machine, "with a couple of lights going across the screen? No, that's not going to help me." But he agreed to give the treatment a go. He knew that he had a problem with patchy memory for the accident. All he could remember at the time was the old man pleading with him to get him out, sketchy details of the ambulance, the police and the air ambulance helicopter, then going to the police station and hearing it come through on the police radio that the man had died.

At first, all the staring at the lights gave him an immense headache. But things soon started to happen. When he was asked to focus on the last image he remembered clearly, the image of the old man's face in the window, drifting in and out of consciousness, he commented, "It's a bit red." Sitha asked him to tell her more. "It's the whisky nose," he said. He had assumed that the redness was from the airbag, but it wasn't. He had been told at the inquest that the old man had been over the legal alcohol limit, but the information did not fit with his current interpretation and so he had not processed it. He knew, as bald facts, that the man had been on his side of the road, and that he had done all he could to avoid him. "It went in one ear and straight out the other. I was hearing things but it was making no sense to me." Now he could connect the facts of the matter to his own intrusive memory, and change his interpretation. As soon as he could understand the events differently, he was able to stop blaming himself.

That became confirmed in the second session of EMDR. Colin remembered the driver of the car behind the Micra, who had stopped to help, saying to him how he had been hanging back for a mile or so before the accident because the old man was weaving in and out of his lane. Colin had not been to blame; the other driver had told him so. The detail fit with the new, emerging interpretation, and made it stronger. So did the memory of seeing the old man's head nodding down, as though searching for something in the footwell, as he drove erratically in Colin's direction. After two EMDR sessions, Colin was no longer blaming himself for the accident. His previous interpretation was in pieces. "I put, like, two and two together," he tells me: the old man's family at the inquest saying that he had had half a bottle of whisky the night before, and the police saying that he had alcohol in his blood. "It just automatically sprung into my mind, why, I've seen that bright red nose somewhere before, and then I realized, a couple of friends I know who were spirit drinkers, they all had bright red noses."

"Before the treatment," I ask, "was the nose not red, or was it red but you didn't notice it?"

"Yeah, it was red but I didn't notice it. I thought it was like off the air-bag, like the bruising off the airbag. How can I put it? Like bruising, basically, off an impact . . . Obviously when you bang your arm it goes red, so I thought it was just off that. And when I was sitting thinking, using the machine, watching the lights go across and putting different little bits of information that I knew but blocked away, and it all came to light, I was thinking, why, it *is* a whisky nose! Bright red nose, bright red cheeks. And little things like that just trigger, I don't know, I can't explain how it happens, it just clicks into place. I mean, if I knew how that machine works, I'd be doing it myself, you know what I mean?"

Colin is no longer a skeptic about EMDR. In his judgment, the treatment has changed his life. At several points in our interview he refers to it as the "magic box." "I don't know how it does it, but it's just bringing different things from the back of my memory. It's brought up a whole different meaning to the way I look at things now." Sitha believes that the intrusive image of the old man pleading to be helped out of the car drove Colin's interpretation that the accident was his fault. It created a filter, a guilt filter, that affected how he was able to process all the other bits of information about the accident. That memory hasn't gone away, but it has been pushed back from the forefront of his consciousness to take its place in a background of other memories. By accessing other memories that live at the same address, the clinical psychologist Kevin Meares told me, the intrusive memory is put back in its place. "It's almost as if you need to breathe life into the memory," he said, "in order for it to become part of the past." The treatment is about filling the memory out, putting detail into it, so that it will make sense as a whole, rather than being just this one extremely distressing fragment. It is still a construction, but a much more balanced and less misleading one.

It is clear that the main issue in Colin's case was not memory distortion, but the integration of memory fragments into a coherent whole that was congruent with his existing interpretation. He blamed himself for the accident, and so his memories reflected that interpretation. When he stopped blaming himself, on the basis of his new interpretation, the memories changed. In other cases, trauma can lead to serious distortions of memory that can be so vivid and compelling that the sufferer cannot believe they are not true. A guilt-ridden survivor of a car crash may be haunted by intrusive images of the scene that seem to show him as having had plenty of time to react and avoid the collision. When the scene is forensically reconstructed through the process of therapy, the patient can have it demonstrated to him that the accident actually happened much too fast, and there was nothing he could have done. Peter's Falklands trauma happened when the study of PTSD was in its infancy, and he was never given the kind of therapy that might have addressed the inconsistencies in his memory (such as the fact that his friend was sometimes still alive as they were burying him). Today's therapists can work with such inconsistencies and make clients examine their memories more scientifically. When the inconsistencies and distortions are exposed, the interpretation can begin to change.

In Colin's case, the image in question was not distorted. Rather, it was a case of aspects of that memory only later becoming noticeable. How was it that Colin could only now be "noticing" that the face in the image was flushed, like a drinker's? The answer is that that detail is only now consistent with his interpretation of events. We remember the past through the lens of the present: what we believe now, what we want now. Memories might be about the past, but they are constructed in the present to suit the needs of the self. Although Sitha finds the language of repression useful as a way of communicating ideas to clients, there is no need to postulate such a process in Colin's case. He did not forget the detail of the face because some unconscious force was driving it from his mind; he forgot it because it did not fit with his current interpretation.

That said, Colin may be wrong about the power of EMDR to effect these changes. There is some evidence that simply asking people to try repeatedly to remember something leads to better recall, but repeated attempts at remembering are also part of standard cognitive behavior therapy (which does not ordinarily involve EMDR). Some have proposed that the movement of the eyes from left to right can improve communication between the two hemispheres of the brain, and that (for right-handers at least) can improve memory performance in lab-based tasks and in studies of autobiographical memory. But it is far from proven that such an effect is at work with EMDR. There is no strong evidence that EMDR causes anything to happen that wouldn't result from regular cognitive behavior therapy, using techniques such as exposure and flooding (in which patients are asked to imagine trauma-related stimuli in the hope that the anxiety response will become extinguished). Richard McNally summarizes it in a rather damning phrase: "What is effective in EMDR is not new, and what is new is not effective."

There is an odd sense in which the traumatized mind is like that of a young child. Small children need to learn to make their way through a landscape of memory, and so must those who have experienced the horrors of war, abuse and disaster. Young children's memory is fragmentary anyway, and even psychologically undamaged children must strive to create coherence just as the traumatized mind must. The difference is that traumatized adults have, before the horror, already established a sense of themselves as extending through time. For the adult trauma sufferer, this can actually add to their problems. Our capacity to suppress traumatic memories seems to become weaker as we get older—hence the particularly vivid experiences of war veterans who, decades later, have cause to recall the horrors to which they were exposed.

Therapy for PTSD centers around this search for coherence. If this integration is not allowed to happen—if the topic of the trauma is taboo, for example, and the events not talked about—traumatic memories remain painfully present at the forefront of consciousness with the potential to burst into color at any time. The goal of therapy is not forgetting, but a different kind of remembering. As Rachel Yehuda has noted, gaps in the memory can be as harmful as the awful memories themselves. The process of therapy is about filling in those gaps and correcting mistaken interpretations, so that the memories can be experienced without the constant effort to avoid them. "Forgetting is no solution," Yehuda writes, "even if the experience itself is painful to remember. These memories are intrinsic parts of people's lives and constitute the essence of who they are."

Colin is living proof of the value of this approach. He has not forgotten the tragic events of that day, and the memory has lost none of
its power to hurt. He still has the occasional sleepless night, and he still
thinks about the man who died. But he accepts that he has responsibilities, to himself and to his family. He is eager to get back to work so that
he can start providing for them again. He takes each day as it comes, and
looks forward to life, where before (he says) he was a "nobody," just going
through the motions. He can catch up with being a dad to his young son.
He realizes that the accident was just one of those things; that he was in
the wrong place at the wrong time. If it hadn't been him, it could have
been the family in the minivan behind him, who were traveling with a
small baby. Every time he feels down he thinks about the baby in the
minivan, and his own baby. He has been lucky: he has had great support
from his family and workmates. He is no longer terrified of sky-blue Nissan Micras; in fact, he thinks he could now get into one of those cars and
drive it. He has been getting back to driving his friend's wagon, practic-
ing on the roads of a nearby industrial estate. A few months later, I hear
word from Sitha that Colin is back at work, after a total of eight sessions
of EMDR. He cannot thank her enough for the help she has given him.
"It's not me," she says, "it's you. I've just helped to bring your memories
back out into the open."

As I write Colin's story, I have a strong urge to visualize the spot
where the accident happened. I feel I should try to imagine his story in as
much detail as I can, so that I can get as close as I can to his memories of
that day. I'm not going to put him through the discomfort of pinpoint-
ing the location himself, so I stay at home and trace the route on Google
Earth, starting at the turnoff from the A1 and clicking through frames
of landscape as I inch along the country road. There is a sudden weird
transition between summer and winter foliage, but on the section where
the tragedy unfolded, the scene has been preserved in springtime. The
images are dated 2009, the year of the accident. I have a weird sense that
I am looking at a frozen part of the past, that I will click forward a frame,

into the next blurry stretch that quickly clarifies, and see the wreckage of the accident: the crumpled blue car, the lorry tipping alarmingly high up on the verge, both facing the same way along the road as though one had been trying to overtake the other. It is a memory that isn't a memory, a flashback for something that I never experienced, but that has become real to me through the imagining. The difference is that I can switch off my flashback. Colin will never be able to forget the events of that day, but now he can hope to remember them differently.

11

THE MARTHA TAPES

"I MEAN, I'VE got a good memory . . . I don't know whether it's manufactured, or it's just hereditary or whatever it is. I suppose you've either got it or you haven't. Is that right?"

My grandmother sits in her green reclining chair, in her usual spot in the corner of her living room. A soft daylight is falling on her from the window to her right, the sill of which is covered with family photographs in neatly arrayed frames. She is wearing a floral blouse and, despite the summer heat, she has a tartan rug pulled up over her knees. I am talking to her about the old days. Hers is a past that stretches back nearly a century; she will be ninety-three this year. She is well used to our routine, the paraphernalia I bring with me, the sight of my laptop and the notebooks I pull out of my rucksack. She doesn't give a second glance to the digital voice recorder I prop up on a little tripod, ready to record what she says for later transcription. It's a long way from my home in the northeast to her flat in Essex, and so we don't meet as often as we would like. When I do get a chance to talk to her, I try to make sure that we make the most of our time together. Although she is in good health, the reality of the human life span means that we are unlikely to be able to keep doing this for many more years, and that unspoken awareness colors every moment we spend together, strengthening through the meeting until it comes to saying good-bye, and I look at that gray sparkle in her eyes knowing that it may be the last time I ever see it.

When she is comfortable in her chair, I set the recording going and let her own thoughts steer the conversation. She is attentive, sitting forward in her chair, her mouth slightly open as though distracted in the process of forming a smile. I might get things started by reminding us what point in the timeline we had reached in our last conversation, but otherwise she makes the decisions about what we will talk about, with only minimal cueing from me. I don't pick her up on details. At first, when she digressed into a topic I hadn't expected, she would worry about whether what she was saying was relevant. Oh yes, I would tell her, it's all relevant. These days she knows that I just want her to talk, as long as it's about the past. Repetition is not an issue. Even if she was aware that she was telling the same anecdote again, I would have no problem with her going over old ground. In a way, part of this is about *wanting* her to repeat herself, so that I can see whether the stories are told the same way each time.

I didn't really know what I was looking for when I first sat down with Nanna, a microphone and my old four-track cassette recorder. We have a large family—Martha has eleven grandchildren and sixteen great-grandchildren—and we have all had a growing sense that her stories need to be set down for future generations. As a child, I would hear tales about Nanna's exotic but only dimly remembered background: the ancestor whose job was to polish the boots of the Tsar's army; the teenage girl (Martha's mother) who sailed back to Lithuania to fetch her two brothers for a new life in England. I wanted to hear Martha's own versions of these stories, and to put them down in a permanent record. For her ninetieth birthday, I presented her with a bound copy of the transcripts that had been made so far, with further copies for the rest of the family. It was a modest gift—what do you buy a lady who has lived for ninety years?—but I knew that it was something that would be treasured by the rest of us, as much as it would serve as a memento for her. As my own interests have evolved and I have become more preoccupied by the processes of memory, the endeavor has changed from a family archiving project into a bigger examination of the mind's grip on the passing of time.

We usually talk for about an hour. I gave up using the cassette recorder after an interview was cut off due to a tape ending, and have switched to a portable digital recorder. Nanna is always interested, responsive and quietly enthusiastic about talking about the past. With her increasing frailness, she doesn't get out much anymore and, while other things become tougher to achieve, this is one activity over which she can exert some mastery. How much her willingness relates to an awareness that she is coming to the end of her life, I don't know. She may recognize that her own memory is failing her, or that it soon will. Or else, like many elderly people, she may simply find herself ready to look back, to make sense of what has gone and reflect on it. For her part, there is probably a curiosity about what I am going to make of it all. It's important to me that I keep her involved in the process, and let her know what she can expect of our collaboration at each stage. Not that she would tolerate anything she wasn't happy with. When she feels that the conversation has gone on long enough, she will abruptly change the topic, and ask when Valerie, my mother, is due to return from the shops, or what I make of the prospects for the weather.

She certainly has plenty to talk about. She was born Martha Weisberg in December 1917, the middle child of two Jewish immigrants. She grew up in the upstairs flat of a house on Hare Street (now Cheshire Street), just off Brick Lane in the East End of London. Her mother was from Lithuania, her father from Russia. There was a close Jewish community in the East End at that time, but Martha went to a mixed school and they were thoroughly integrated with the Gentile families around them.

Her father, Abraham, had worked in a cigarette factory, but when the Depression closed the factories, he had to set himself up in business, selling bagels from a stall on Brick Lane. It was a hard living and, like many of the families around them, they were not well off. Martha's mother, Frieda, was a bright, resilient woman who picked up English pretty well, unlike her husband, who never mastered any language other than Yiddish. When Martha had learned to read at school, she passed on the skill to her mother, who could already read the Hebrew prayer books and the newspapers printed in Yiddish but had never learned to read the language of her adopted country. Nanna thinks she must have been five or six when she taught her mother to read English. Frieda worked as a tailoress, like many young Jewish women, and she would also take in buttonhole work that would occupy her in the evenings once the children were in bed. Although they were family stories rather than her own memories, the young Martha realized that her mother had had an extraordinarily tough life, having been frequently uprooted from her latest temporary home to escape the violence of pogrom-torn Lithuania. At eleven, she had crossed the Baltic to begin a new life with her uncle in Birmingham, and at fifteen she had gone back, alone, on the boat to Lithuania to bring her two remaining brothers to England.

Fascinating as the family story is, I am really here to listen to Martha's own memories. The fact that it is me, her grandson, asking the questions might open a particular path into that autobiography. She sees her four children (including my mum) from day to day, and they probably talk about day-to-day things. My visits are rarer, and my scarceness may prompt her to approach the task differently. Perhaps the gap of two generations makes her feel that she can speak more freely. I have often had the feeling that she is sharing some of these memories for the first time. Things are coming to light that the rest of my family have not heard before. Although her children have always asked her about the past, I suspect that I go about questioning her in a particular way, and so she might tell me things that she hasn't told others: not because she wanted them kept secret, but simply because they didn't come to mind. One thing I have learned about remembering is that it is a social process: it happens in collaboration with other people. Her memories with me may not be the same as her memories with others. The reality of remembering is always contextual.

It's not just that I am a different partner in her reminiscing about the past. I also have a different agenda. Listening back to the recordings, I can see that my style is to focus on certain feelings and impressions: the sight of her father, Abraham, setting off to work at his bagel stall outside a fish-and-chip shop on Brick Lane; the waves of fear transformed to relief as the bombs dropped around her during the Blitz. I want to know what it was like to grow up in the Jewish East End under the shadow of Nazism. I want to know whether she was aware of the tailoring work her mother took in, whether she lay awake listening to the swish of the fabric and the whispered Yiddish conversations about money. I want to know what went through her mind as well as her body when she stood outside in the freezing cold of dawn, minding the bagel stall. Her mother suffered from rheumatism, which is why Martha used to offer to do the work out in the stall in the cold. Generous as the gesture was, there is room in her memory for a slight resentment of her elder brother for not taking his turn. In reconstructing her childhood with her, I don't just want to know the facts of what she did. I want to understand her motivations, so that I can know who she was as a person, and try to work out the younger Martha's relationship with the person she is now. Memory narrativizes us; it turns us into characters in a novel. It makes motives and context matter. What we remember is shaped by the people we were then—not just what happened to us, but what kind of individuals we were—as well as by the people we are now.

And there is much more "then" than there is "now." It is a cliché that the elderly are stuck in the past, able to remember events clearly from decades before but amnesic about what happened days or even hours ago. We have already seen how events from our late teens and early twenties (the so-called reminiscence bump) stick in memory better than anything else. Nanna's testimonies sample this privileged period for remembering very thoroughly, confirming the fact that the reminiscence effect is detected most strongly in people older than about sixty (and is even visible in elderly people with Alzheimer's). When I ask Martha about the past, she talks about the 1930s, not the 1980s. She can tell me about meeting my grandfather at the Labor League of Youth, and of how they negotiated the problem of her "marrying out" (Bill was a Gentile). These are events from seventy years ago, and they are much closer to the surface of her consciousness than the events of a few months ago.

In a 2009 interview with the *Guardian*, the novelist Penelope Lively described how, with advancing age, she had become more conscious of memory's ability to let us access the past on demand. "In old age, you realize that while you're divided from your youth by decades, you can close your eyes and summon it at will . . . The idea that memory is linear is nonsense. What we have in our heads is a collection of frames." All time periods coexist in the aging mind, and the calendar is not a good guide to them. As the heroine of one of Lively's novels puts it, "There is no chronology inside my head."

Memory's obsessive return to young adulthood is not just nostalgia for a vanished past. Rather, the reminiscence bump seems to be a basic fact about how autobiographical memory works, and cognitive scientists have explored different possible explanations for it. One idea is that your youth is remembered better simply because that's when the momentous events in your life will have tended to have happened. Big events are more salient and important to the self, and salient things are remembered better. For Nanna, the great upheavals mostly happened before the Second World War, and so that's where the story of her life has its focus. Confirming this "big events" view, studies have shown that many events from the reminiscence bump involve doing things for the first time. Another possible explanation for the reminiscence effect is that the brain is simply better at encoding stuff when it is younger, and so more details get laid down. This seems unlikely, though, because the basic machinery of autobiographical encoding is probably at its most powerful in middle childhood, which would mean that the reminiscence bump should be occurring quite a bit earlier.

A third explanation is that the events that happen in the privileged period of the reminiscence bump were more important for the rememberer's particular life story and in molding the person she turned into. When the events of Martha's testimony were happening, they were shaping the person that she became. They left their mark, in the way that events from the 1980s and 1990s didn't. If it is true that the big, self-shaping things happen to you in early adulthood, then this information should be reflected in cultural wisdom about the human life span. To test this idea, Dorthe Berntsen and her Danish colleagues recently asked children aged ten to fourteen to write narratives imagining the lives ahead of them. When these future life stories were coded, most of the imagined events clustered in young adulthood, like the rites of passage of getting married or moving into one's own house. The children could not have been favoring events in young adulthood because of any superiority of encoding in that period, because young adulthood had not yet happened to them. As a control, the researchers also asked the children to generate future events on the basis of simple word cues. In this instance, the events generated were not structured by cultural wisdom about what, in a human life, is supposed to happen when, and there was no peak in young adulthood.

Another cliché about memory in old age is that time seems to go past more quickly as life goes on. This may just reflect the workings of the reminiscence effect: as you get older, the big stuff happened further back in the past, and so the present seems relatively less full of salient events. When you then think back to what you've done in the last year, say, less comes to mind, because there really is less to distinguish those months. Douwe Draaisma attributes an early version of this idea to the nineteenth-century French philosopher and psychologist Jean-Marie

Guyau. "[T]he impressions of youth," Guyau wrote, "are vivid, fresh and numerous, so that the years are distinguished in thousands of ways and the young man looks back on the previous year as a long sequence of scenes in space." Later in life, there is less to distinguish the passing moments. In William James's mournful phrase, "The days and the weeks smooth themselves out in recollection to contentless units, and the years grow hollow and collapse."

To make this argument convincingly, you would have to say that your subjective experience of the present is dependent on your rate of laying-down and recall of memories. The neuroscientist and writer David Eagleman has studied the relation between time perception and memory, and has argued that life goes past more slowly when you are young because you are encountering more novel information, and thus encoding new memories at a faster rate than in your more mature years. When you stop to judge the length of a period of time that has elapsed, it seems more action packed, and so you assume that it must have gone by more slowly. In contrast, the years of your dotage give your brain fewer novel experiences to process (partly because much of what you do is governed by familiar routines), and so the time is judged as passing more quickly. In Joshua Foer's phrase, "Monotony collapses time; novelty unfolds it." If this is true, then by asking older people to pay more attention to the events going on around them, you should be able to slow time down for them. There is not yet any strong scientific evidence to support the idea, but it seems nicely possible that you might slow down the onrush of the years by stopping every now and then to smell the roses.

It is likely that other factors are involved in the life-speeding effect. One possibility is that when you are younger, the same amount of time is a much bigger proportion of your life. Another idea is that the apparent speeding up of life corresponds to a slowing down of basic metabolic processes. Our bodies are regulated by all sorts of biological rhythms, and it's likely that these can act as regulators of judgments of time. It is certainly true that the elderly show deficits in time perception. If you ask a group of older participants to close their eyes and wait until a minute has passed, you will find that they routinely overestimate the time period. Younger adults are much more accurate, although young children err the other way, and say that the allotted time has elapsed sooner than they should.

It is likely, then, that Martha's biological clock is running more slowly than it did when she was younger, and this might combine with the reminiscence effect to make the intervening decades feel as though they have passed in a blur. "The years certainly have flown by," she says on more than one occasion. Sometimes, though, I get a sense that she's not so much misjudging time as standing outside it. Her ability to move between time frames underlines Penelope Lively's point about memory being achronological in old age. Another novelist, Hilary Mantel, observes that memory is like "a great plain, a steppe, where all the memories are laid side by side, at the same depth, like seeds under the soil." On another occasion Mantel expands: "In our brains, past and present co-exist; they occupy, as it were, adjoining rooms . . ."

Keeping track of the time frame you currently inhabit, while traveling at will to another time frame that might be several decades distant, is a considerable feat for an aging brain to pull off. That's why you can sometimes get a sense, when talking to the elderly, that they are reliving the past for real. Nanna will be talking about her feelings of anxiety during the Blitz, the uncertainty about where the bombs are going to fall next, then will suddenly jump forward to the 1970s, to describe the traveling adventures of her youngest son, Philip. I tell the rest of the family that we need to be patient with these slippages in time. They are not signs of dementia, but simply the struggles of an aging brain to perform a difficult juggling act with its multiple selves.

Time has effected many changes on my grandmother's brain, just as it has on her body. She has lost five inches in height since she was a young woman, and also a proportion of her gray matter. It is not that the cells in her cortex have been dying, but more that the synaptic connections between them have become less dense. The normal aging process sees slight reductions in the volume of medial temporal lobe structures such as the hippocampus. A bigger deterioration is seen in the prefrontal cortex, so crucial for memory retrieval and source monitoring. The elderly seem to have particular difficulties with monitoring the source of their memories, leading them to be especially susceptible to false remembering. Several studies have shown that older adults find it hard to keep track of the source of the information that they remember, such as which of two speakers told it to them. In one famous example, Ronald Reagan repeatedly told a moving story during the 1980 presidential campaign about a pilot's heroic attempts to evacuate a stricken bomber. What he thought was a genuine event actually turned out to be the storyline from a Hollywood war movie. It may not have convinced the sniggering journalists who uncovered the error, but it was a natural result of age-related memory decline.

In one recent study, Jon Simons and his colleagues at Harvard wanted to find out just how much information about source elderly participants could remember. They may not remember vivid details about which speaker had provided the material, but would they remember partial contextual information such as the gender of the speaker? In many situations, having that partial information about source at your fingertips might be enough to make an accurate judgment. President Reagan's blushes might have been spared, for example, if he had been able to remember that he was recalling a movie plot rather than the utterance of a real person, even if he couldn't remember the exact movie.

To test this idea, the researchers compared the source memory performance of a group of sixty- and seventy-year-olds with a comparison group of younger adults. Recordings were made of four different speakers (two men and two women) reading out a large number of trivia sentences. The sentences were chosen so that the average person would not be able to say whether they were true or false. For example, one statement read "Al Capone's business card said he was a used furniture dealer." The participants' task was to judge whether the speaker believed that the statement was true (they were not told that they would later be tested on their memory). Participants heard the statements read out through headphones and also read them for themselves on a computer screen. When the statements had been studied, the volunteers were given a surprise memory test of the original items, plus a few new ones. They had to indicate whether they had heard each statement before, and if so, which of the four speakers had generated it.

It was clear that the older adults were not as good at identifying the correct source of the statements. When the responses were analyzed according to their use of partial source information (such as whether the subject got the gender of the speaker right, if not their exact identity), the older group still showed a deficit compared to the younger adults. In a condition where the older adults were given three chances to study each statement, instead of just one, they caught up with the younger adults on both specific-source and partial-source memory. The researchers concluded that aging affects your memory for specific sources, but it interferes to a similar extent with your ability to use bits of contextual information in helping you to narrow down sources you can't remember outright.

Daniel Schacter has pointed to some of the real-life implications of this loosening grip on source memory. Some older people get reputations as incorrigible gossips. Keeping track of which bits of information are supposed to be secret is of course a challenge for source memory: to know that it's a secret, you have to know who told it to you, with whom the contract of secrecy was made. So it's no surprise that older people are less good at it. In our family, over the years, there have been plenty of news items that were supposed to be kept quiet, for a little while longer at least. Quite unwittingly, Nanna has often been the first to break these confidences. For example, I have more than once heard the first news of a pregnancy from her, and then heard her immediately follow it up with, "Oh, perhaps I wasn't supposed to say that yet." She doesn't engage in more mischievous gossip, not with me at least, since family is our only overlapping social group. But Schacter's experiments show that older people really do have trouble remembering which bits of information are meant to be confidential. "This finding does not necessarily mean that you should never trust your grandmother with a secret," he writes, "but you should probably handle such matters with care."

One thing I am curious to ask Martha about is the quality of her memories. An age-related decline in prefrontal cortex functioning should affect those interactions with the medial temporal lobe system that are so essential in constructing a memory, and thus perhaps the rememberer's ability to integrate different contextual features. This would account for findings that our memories become less vivid as we get older, with fewer perceptual and background details. It is also one reason why my questioning style focuses on the immediacy of Martha's memories. Memory researchers recognize the importance of separating out different components to the quality of our memories. For example, you could have a memory with lots of perceptual detail that nevertheless wasn't very life-like, or you could have a vivid memory that didn't really transport you back to the moment in a traveling-through-time kind of way. The British study of nonbelieved memories, for instance, found that such memories were equivalent to true memories on visual qualities and mental time travel, but were generally less vivid.

I ask Nanna whether, when she casts back into the past, she relives her experiences as if they were happening to her again. Do those events from long ago come back to her clearly and colorfully, or are the memories sometimes a bit fuzzy? She replies by pointing out that she is ninety-three and that she has "a good few years to remember." Then she tells me that she has a good memory, and wonders whether that is a hereditary, natural thing. (I don't have a good answer for her.) I'm not sure that she really understands the point about vividness and mental time travel. Simply going on the descriptions of her memories, I doubt that they have quite the same lucid quality that a younger person's memories would have. This fits with findings that semantic memory is preserved (or even enhanced) in old age while aspects of episodic memory deteriorate. Martha doesn't often tell me about detailed subjective impressions or any but the most general emotions. When I ask her to conjure up an image of her father setting off for work at his bagel stall, she can recall that he wore a suit, and that there may have been a trilby to go with it later in life. But the image is not detailed. She is telling me the facts, not the impressions.

All in all, then, I have my doubts that Nanna is really traveling back in time in some of these memories. This is not the same question as whether she sometimes gets confused between time frames. Autonoetic consciousness is about having the capacity to make an active choice to go back and reinhabit a past moment. Martha could be getting confused about time frames and still have good abilities to travel back in time to relive a particular event. To go with the reduced contextual detail of their autobiographical memories, the elderly have also been shown to have difficulties in richly imagining future occurrences. In one recent study, researchers asked a group of sixteen older adults, with an average age of seventy-two, to generate past and future events in response to cue words. The resulting narratives were coded for episodic and nonepisodic (or semantic) details. Confirming previous findings, the older adults produced fewer episodic details and more nonepisodic details (relative to a comparison group of adults in their twenties) for past events. But the same pattern was also shown for the future events. Given Martha's very advanced age, it would seem insensitive to ask her in too much detail about the future (for the same reason, the older participants in the study described were not asked to think more than a few years into the future). If I did ask her, I suspect that I would find that her imaginings were factual rather than richly contextual. She would tell me what might happen, but not necessarily how it would feel.

Perhaps her memories for mundane events are not particularly vivid, but what about the momentous events in her life? One kind of memory that is particularly rich in subjective detail is flashbulb memories. If I cue her with historic events and ask her to recall what she was doing at the time that she heard news of them, will she show the flashbulb effect?

The answer to this question may depend on how long ago the momentous events happened. For new events, it is likely that any flashbulb memories will be fairly short-lived. The researchers who asked students how they learned about the resignation of Margaret Thatcher asked the same questions of some elderly people. All were able to give detailed accounts of how they came to hear of the event. A year later, however, only 42 percent of the elderly people's memories of the event met the criteria for a flashbulb memory (defined as a memory that stayed very consistent over the time interval), compared with 90 percent of the memories of the younger participants. Findings from other studies confirm the conclusion: when it comes to forming new memories, the flashbulb effect is not as strong as it is in younger people.

This is not to say that Martha could not have flashbulb memories for events that happened when she was young, such as her marriage proposal or the outbreak of war. In a groundbreaking study by Dorthe Berntsen, elderly Danes were asked about their memories of the invasion (in 1940) and liberation (in 1945) of Denmark. Responding to a questionnaire, each participant had to say if they remembered where they were and what they were doing at the time of the events, and (if they indicated that they did remember) to provide further details of their personal context. They were also asked to provide detailed descriptions of their most positive and most negative memories of the war. They were then asked to rate their own memories on measures of vividness, mental time travel, re-experiencing, perceptual detail, and so on. Specific flashbulb qualities were recorded by scoring the memory reports for details about ongoing activity (what the participant was doing at the time of the event), information source (how they learned the news), the emotional responses of those and others around them, the presence of other people, and aftermath (what happened immediately after the news broke).

Almost all of these elderly participants had flashbulb memories for the two events. Archival records, such as weather reports, were used to confirm the veracity of the memories. Those memories turned out to be rather accurate, with most participants, for example, correctly describing the weather on the days of the events. Compared to a control group who were not born or were very small at the time of the events, many of the older subjects gave very accurate answers about the time of the invasion. The researchers were also able to examine the effect of personal emotional salience on the vividness and accuracy of the gathered memories, by taking advantage of the fact that some of their sample had been involved with the resistance movement. The narratives produced by this subsample were more detailed, lifelike and accurate than those produced by those who did not report any connections with the resistance, suggesting that when there is a very strong personal relevance (intense personal danger), flashbulb memories can be even more powerful.

The findings from the Danish study show that flashbulb memories laid down early in life can persist over very long periods. When, in one of our earliest interviews, Martha was asked about her memories of the war, she recalled a few events in some detail:

> *The only thing I can recall, vividly, was, um, Bill and I went . . . to the cinema one night, in Loughton . . . and as the program, as the film was finishing, there was a terrific bang, and of course everybody, you know, we'd never encountered any bombs, because it was the Phoney War, and then when it was, I think, the end of the film, and we stepped out the cinema and we could smell this cordite, and a bomb had dropped just up the road.*

A little while later in the same interview she recalled watching the bombs drop over London from her house in the suburb of Buckhurst Hill:

> *Bill used to go to the top floor and he used to look out and see the bombs dropping, and telling me about it. One particular night, um, it was a Saturday night, and we were also quite close to a balloon barrage station, where there were the people serving, there*

were lots of recruits there, and one particular night, which was a Saturday night, Bill was standing up there, he used to have the window open, and he saw a great big flash, and he knew something had happened there, and it was a tragedy because about a hundred and twenty of the militia were killed that night because they were all, you know, all the youngsters, they were all in the pub . . . There was, in Chigwell, you see, we were quite close to this balloon barrage station, we could walk to it, and that was Chigwell, and there were just open fields straight across, Bill could see, straight across, and he saw this terrific flash . . . And that was, er, that was a landmine . . . Which, you know, they came down, you didn't hear it, you might have heard the swish of the, the parachute . . . And that was dropped on the pub, that came down on the pub, so there was no warning and there were about a hundred and twenty young recruits in there, being a Saturday night.

These were clearly frightening events, fertile ground for flashbulb memories. For Martha, the permanent memories are not of sitting around the wireless listening to Chamberlain's announcement in September 1939, but of seeing the first bombs drop on London:

[I]t was the Phoney War, and nothing happened, and then one, one day, I mean I'm going back now, one day Bill and I were just, it was a lovely, well it must have been August, end of August, September, and we were just having a walk, the sky was blue, and we thought we'd just have a walk, we used to walk up to the balloon barrage, you know, it was a nice country walk. And we had to sort of cross the bridge, the railway bridge, 'cause I told you we lived near the railway, and there was a warden there, and he said to us, "If I were you, I'd go, I'd go back home," he said, um . . . the warnings have, I don't know if the warning had, had gone when we left home, it was only just round the corner, but he said they're fighting in the skies,

which, well, it was very high, which if you looked up you could see them, he said, "I would advise you to go home," which we promptly did . . . and then . . . the all-clear didn't go . . . didn't go . . . it got towards evening, it was very unusual for it to, the warning to go on for so long . . . So we couldn't understand why it, it was, it had lasted so long. It got to evening, and of course it doesn't get dark too early, and you could see this red glow, and that was the beginning of the Blitz.

We have seen that flashbulb memories may have special properties because emotional arousal at the time of the event triggers amygdala activation, in turn affecting protein synthesis in the hippocampus. I have no doubt that Martha has felt emotions as intensely as any of us; she has probably been exposed to far more frightening things than most. But it may be that she is from a generation who didn't talk about emotions quite as much as we do these days. Generally, not talking about the past will probably have affected the way she can talk about it now, as the research with parent-child conversations about memory has shown. Oddly, though, the elderly have been shown to be particularly good at using the emotional texture of an event as a source cue. Older people's memory narratives of fictional events (when they are asked to create them for experimental purposes) contain more elaborations on personal thoughts and feelings, compared to younger adults'. In fact, selective attention to positive emotional memories has been highlighted as a feature of memory in old age, one of the functions that, along with recognition ability and automatic memory, seem to remain intact throughout life.

I also wonder whether I am simply noticing a difference in Martha's style of testimony compared to that of a younger person. It's possible that she is reliving the experience in the most vivid possible way, and yet her style of retelling is not as subjective and first-person as a younger person's might prove to be. For comparison, I ask Mum (who herself is approaching seventy) about her memory of the death of Frieda, her grandmother, in 1952. She recalls a vibrant, perceptually rich experience of sitting in the back of the family car at Gants Hill, watching her mother emerge against the lights of the tube station in a pale tan coat and long full skirt, and climbing into the front seat with tears gleaming on her cheeks. She is clearly back there in the moment, in a way that Martha doesn't quite seem to be. Perhaps Nanna has certain ideas, specific to her generation, about the kind of things you are supposed to say about your memories. Perhaps, like a nineteenth-century novelist, she narrativizes herself in the third person more than the first person. Either way, her memory narratives do not generally have quite the same vividness that you might find in someone even twenty years younger. They are schematic, like well-rehearsed stories, told with slightly different emphases, and sometimes details, each time.

IF CONSTRUCTING AN AUTOBIOGRAPHICAL MEMORY is a struggle against the centrifugal forces that would otherwise separate out its elements, then it is easy to see how narrative provides a helpful structure. When you tell a story, the details of context, setting, characters and their motivations are fixed into the tapestry. If you can keep a grip on the narrative, you get these disparate details for free. Ironically, this may also make memory less susceptible to some of the reconstructive errors that dog ordinary remembering. The worse your memory gets, the more protected it becomes from some of memory's foibles. If we choose not to rely on narrative, we have to weave the tapestry anew each time, and that's where the errors creep in. Although other factors conspire to increase the fallibility of Martha's memory—source memory deficits due to pressure on her weakening prefrontal cortex, for example—an increasing reliance on narrative structures gives her tales about the past (like those of the amnesiac Claire) a particular consistency and authority.

Memory is Nanna's autobiography, her life story, and it has a core of certain events, mostly from the period before the end of the war. These are facts she will not lose track of, time frames she gravitates back to. The stories of teaching her mother to read English, her courtship with Bill, the anti-fascist marches and the outbreak of war have been told many times before, and the manner of their telling acquires a permanence, even if the original information from which they were constructed may not always have been completely accurate. This may be the only time in our lives when the "mental DVD" analogy becomes anything like accurate.

I look for repeated themes in her memories of the past, and find a few stories that come up more than once. On three different occasions, for example, she tells me how her husband, Bill, told her that she must give up her job when the expected declaration of war came, and of his displeasure when she broke the news to him that she had actually gone and done so. On other occasions, the authority of her storytelling can be misleading. For example, she tells the story of a stampede in the Bethnal Green underground, as people rushed to take shelter from an air raid. Listening to the vividness of her description, and comparing it to her other memory narratives, you might be surprised to realize that she wasn't actually there at the event.

The consistency of her memory narratives makes me wonder whether there is any scope for her life story to continue to develop. As she doesn't get out so much now, new adventures are thin on the ground. But could new details come into her narratives of the past? Could her *feelings* about the past change? I know that a change in emotion can unlock inaccessible details from memory. In his 2011 Man Booker Prize–winning novel, *The Sense of an Ending*, Julian Barnes describes how a shift in his protagonist's feelings toward his former lover's parents unlocks new memories of their relationship. "But what if, even at a late stage, your emotions relating to those long-ago events and people change? . . . I don't know if there's a scientific explanation for this . . . All I can say is that it happened, and that it astonished me." All this talk about her mother, for example, might have made Martha come to feel differently about the difficult life that Frieda led: more sympathetic, perhaps, more able to identify with the immigrant seamstress working long into the night as her children slept. Might new memories follow any such shift in Martha's feelings?

I have been on the lookout, then, for the uncovering of "new" memories, moments of experience buried for many years and only just now emerging into the light. She still has the capacity to come up with completely new information, such as in her tale of a suitor called Willy who was soft on her in the years after Grandad's death in the late 1970s. But the cognitive challenge of retrieving a "new" autobiographical memory may depend on her receiving the right cues. Not everything that has happened to her has been encoded in memory, and those details that have been encoded may require certain conditions to be in place before they come back into consciousness.

This made me recently start to wonder whether, if I changed the format of the interview, we might together coax out some new stories. One fundamental principle of memory, encoding specificity, means that information is recalled better if it is recalled in the same context as it was learned. In the case of someone born into an immigrant community, a crucial contextual factor is language. I had been reading a study of memory in a sample of twenty young Russian immigrants to the United States who had left their home country in their teens. The participants recalled more memories from their childhoods when they were interviewed in Russian, compared to when the interview language was English. In a second experiment, the researchers independently manipulated the language of the memory cues and the language of the interview (for example, one condition might involve English cue words embedded in an interview that was otherwise conducted in Russian). They concluded that the interview language and cue language made separate contributions to the unlocking of autobiographical memories, such that the effect was strongest when both the cues and the ambient language matched the memories.

Similar findings have been reported in a number of other studies. Matching the language of recall to the language spoken when the events happened seems to free up otherwise inaccessible memories. We have already seen that language plays a vital role in mediating our autobiographical memories. Although we talk to ourselves about our pasts, most of that language use happens in talking about the past with other people. I took my cue from several mentions Nanna had made of the fact that her father did not speak English very well. Instead, he and Frieda communicated mainly in Yiddish, which would have been the language that the young Martha would have heard at home. Would her memories pan out differently if she was interviewed in Yiddish rather than in English? With Nanna's blessing on the experiment, I decided to try to find out.

There followed a quest to find a Yiddish speaker who could travel to Essex to interview her. Many of the people I contacted were themselves elderly and unable to travel. Through a connection at the Department of Hebrew and Jewish Studies at University College London, I was put in touch with Sima, a Lithuanian Jew who is now settled in the UK. Sima agreed to travel to Essex and talk to Martha in Yiddish, while Mum and I listened.

I had very little idea how the experiment would work out. Although she had not been exposed to the language for more than half a century, Martha was a willing participant. For my part, I first had to ask whether she would remember anything of the language that I assumed would be associated with her earliest memories. Research shows a great deal of variation in how well people retain their mastery of a language spoken decades earlier. Some forget their native tongue completely, while others seem to revert to it naturally as they get old. This latter phenomenon is known as *language reversion*, and it may result from the second language becoming forgotten while the speaker's grasp of the first language strengthens with age. I didn't really expect Nanna to be able to form any sentences in Yiddish, but I thought she might recognize a few words, and that she would understand at least some of what was said to her.

Sima would also add another element to our project: she was a complete stranger to Martha, and so would create an entirely new social context for her remembering. Sima had seen some of the earlier transcripts and was prepared to ask some of the same questions. I also hoped that, as someone who had conducted research into the prewar East End Jewish culture, Sima would be able to introduce some redolent cultural specificities that might trigger new memories. I wondered whether I might even witness Nanna reacting emotionally to the experience, as she realized that she was remembering an event for the first time.

We agreed that Sima would start off speaking in Yiddish, and then translate her own words if Nanna was struggling to understand. Nanna sat in her usual chair, her mouth open in an attentive, bright expression. She responded to the standard greeting, "*Sholem aleykhem*," without any need for translation. She understood several of Sima's subsequent questions, answering confidently in English. She recognized the word *shadkhen*, or marriage broker, recalling how her parents' marriage had been arranged. She remembered the Yiddish names for the baked goods sold by her father in the bagel stall. The food of her youth seemed to provide a particularly strong connection to the old language: *tsimes, lokshn kugl, shabbes challe*, and of course *gefilte fish*. But many of the questions were beyond her, and Sima had to translate them into English before Nanna could respond. As she leaned forward in her chair, I wondered whether she was struggling to hear. But she understood the English with no problem, so I was probably only witnessing her battle to understand the unfamiliar language.

In many ways it was a typical interview. Nanna told Sima some of the details she had previously told me, such as how she had helped out with the bagel stall on freezing cold mornings before school. As thoughts tailed off, her statements would generalize into soulful affirmations of the good fortune she had had in life. As with our previous interviews, she would sometimes jump around in time, suddenly fast-forwarding to the 1970s before returning to her previous thread. Contrary to my secret hope, there were no sudden outpourings of new memories. As I listened to the language that had been part of my own youth (I will still accuse a mischievous child of *shmeikhlt vi a vantz*—grinning like a bedbug—

echoing the phrase I used to hear as a child), I realized that Nanna had probably not spoken as much Yiddish at home as I had previously assumed. It was her parents' language, used for grown-up conversations from which children were presumably often excluded. When she could, the young Martha spoke English: to her friends, to her mother, to her brothers, to herself.

As the interview wound down, I switched off the recorder and went into another room to phone for a taxi for Sima. A moment later I heard the chatter of an excited conversation. When I went back into the room, Mum told me that Nanna had suddenly recalled the place in Lithuania from which her mother had emigrated. I had quizzed Nanna on this many times, and she had never been able to remember any details. But as soon as Sima mentioned that her own family was from Kovno, Nanna pointed out that this had also been her mother's hometown. Mum and I were amazed. "I always told people it was Kovno," Nanna insisted, "whenever they seemed interested." Afterward it transpired that Nanna had occasionally mentioned her mother's Kovno origins to her friends in Hackney, before she got married and moved away. She hadn't actually forgotten the detail I had been reaching for; she just hadn't remembered it for seventy years.

THE DAY AFTER MY NEWSPAPER piece on Martha's interview appears, I receive an email from a woman in Essex who says that her mother has just read the article and recognizes Martha as an old classmate. Sadie had been in the year above Martha at Mansford Street Central School in Hackney, and recognized her from the description of my great-grandfather Abraham, the bagel seller. Sadie is now ninety-four and still quite mobile, and so we arrange for her daughter, Hazel, to bring her to the nursing home where Nanna has been staying since she broke her wrist in a fall at her flat in January. Unwilling to miss what promises to be a unique reunion, I drive down to listen in, having gotten permission

from all concerned to record the interview. I sit with Martha in the big
sitting room at the end of the corridor and watch a white-haired lady
come into the room supported by a walking stick. Martha is sitting in a
wheelchair and cannot see Sadie as she first comes in, but I watch a smile
break over her face as she hears her old friend's voice, and the two ladies
fall to talking: first about the niceties of home and family, and then, without much prompting, about the past.

It is extraordinary watching them reminisce together. They begin
with factual details—Martha asking Sadie when she left Leytonstone, her
home on the border between Essex and east London—which quickly trigger genuine memories of how Martha and her friends went to visit Sadie
there at the family shop. Sadie remembers that friends of hers from school
made the trip on Sundays, and that they would go walking and picnick-
ing in nearby Snaresbrook. Martha's comment brought the detail back to
her. "I've got an awful memory," Nanna says in reply. "I remember things
I shouldn't, and things I should . . . you know?" She expresses surprise
that Sadie has been in her current flat for forty-five years: "I thought it
might have been a recent move." I have the strong sense, both now and
later, that Sadie's presence has rewound time for Nanna. She is dismissive
of her current incapacitation and of the "rules and regulations" of life
in a nursing home. On the phone a few days ago, she told Sadie that she
lived in a flat in Buckhurst Hill, which confused Sadie and Hazel when,
for this meeting, they were directed to a nursing home near Chelmsford.
Martha and the family moved out of the Buckhurst Hill flat in 1949, and
Sadie left the shop in Leytonstone in 1943. Keeping up with the various
further moves that have taken place over the years is always going to be
a struggle.

The details that Nanna wants to tell Sadie are exactly those details
that have featured in our own conversations. Facts about when and how
she got married, how the outbreak of war affected them, how they sur-
vived the bombing, when the children came along and where they lived:
these are cornerstones of Martha's life story. Both ladies keep returning

to affirming, heartwarming details of their children, grandchildren and great-grandchildren, which are interspersed with some general comments from Martha about the unlikelihood of their meeting and the passing of time. "It's a small old world," she says more than once; "it's amazing how the years go by."

Then Sadie asks, "Can you remember the old school?" She recalls how she took her own children back there in 1990 to have a look around and see if the place had changed. This relatively recent memory provides a new focus for their joint reminiscing. They talk about how the girls would set up stall in the "housewifery" department and sell rock cakes to the boys. On Fridays the school used to knock off early for the Sabbath. Mansford Street wasn't a Jewish school, but so many Jewish children went there that the "double session" became an established feature of the schedule. They would have a shortened lunch break and end school at half past two. We look at an age-darkened class photo from 1929, taken when the children were eleven or twelve, and many of the names come back to them. Sadie's visit from twenty-one years ago is relatively fresh in her mind (she recalls finding her husband's initials still carved into one of the desks, many decades on), so she is the one quizzing Nanna about the name of the headmaster, which Martha remembers immediately: "Mr. Hawker." He used to start the school day with bird impressions. "That was our assembly," Sadie says, and laughs. Sadie doesn't appear in the photo, as her father was critically ill in the hospital at the time (he was a French polisher, and the chemicals had ruined his health) and she was taken out of school. One girl in the photo, whom Sadie remembers as Eliza West, had a very distinctive 1920s hairstyle, a flapper-girl bob with a fringe. They remember that Eliza was ill at the time and died shortly after the picture was taken, at the age of twelve. Sadie went to the family's house with a wreath and saw the dead girl laid out. "It was the first time I'd ever seen anybody dead," she says. "I've never forgotten it. It stopped me being afraid."

After they left school, Nanna used to go over and visit Sadie in her family's haberdashery shop. She and a few other girls would take the tram from Hackney to Leytonstone on a Sunday, and Sadie's mother would pack them sandwiches to take out to the duck pond in Snaresbrook. Sadie lives near there now, and she remembers as a schoolgirl, perhaps even on one of Martha's visits, seeing the building that is now her home being built (it had a distinctive green roof, and she had never seen a green roof before). The ladies work out that those visits must have ended around 1932, so it has been nearly eighty years since they met.

At one point Hazel, Sadie's daughter, asks whether the two ladies recognize each other. "I don't know," Martha replies, to which Sadie responds, "I recognize your *daughter*." Mum, who is watching the proceedings with Hazel, does indeed look a bit like Martha did as a younger woman. It seems to make sense that Sadie has held an image in her mind of Martha as a much younger person, although it is also strange that she feels a stronger sense of recognition for her friend's daughter than for the person herself. I suspect that our images of ourselves and those close to us don't quite keep up with the reality of aging, especially when the reunion happens after such a long stretch of time. In the search image she has constructed, Sadie is looking for the Martha she knew back then, rather than the frail old lady she is now. Both women comment on the fact that they are each at least five inches shorter than they were in their heyday. There is also a disagreement about Martha's hair color: Sadie remembers her as having been quite fair, while Mum thought she had always been dark. Martha herself confirms that she used to be fairer, although she thinks there may have been a contrast effect as many of the Jewish children around them were very dark.

Sadie clearly remembers the bagel stall, and we spend some time trying to pin down its exact location, at the junction between Osborn Street and Brick Lane. Sadie recalls how the other children felt sorry for Martha for having to stand out in the cold manning the stall, from six in the morning until it was time for school. Martha says that she did this partly to protect her own mother from having to do the same. Sadie recalls that there was a rival bagel seller a few doors farther up the road, a grumpy woman who used to shout, "Bagels, three a penny, three a penny," and then curse people if they didn't buy them. We speculate on which synagogue Abraham would have worshipped at, and agree that he would have had a good selection to choose from. Sadie recalls being a bridesmaid at the wedding of a cousin who lived on Fournier Street, right across from the *shul* (now the Jamme Masjid mosque). Arranging for a taxi would have been absurd, and so they laid a red carpet across the street from the doorstep of the house to the threshold of the synagogue, so that the bride could walk to her wedding in style. Both women remember the Jewish and Christian communities as being very well integrated. On another occasion, Martha told me how she and Bill joined the anti-fascist demonstrations at Cable Street, but in fact those racial tensions were arising later, in the run-up to the war. It was the Christian families, if any, who faced social problems. Martha recalls the black eyes among the female customers on Monday mornings, and walking past East End pubs and seeing barefoot children sitting crying on the pavement, while their parents drank the evening away inside.

In previous interviews, Martha had commented to me about how her memory works better when it is cued. "If you said to me a certain incident, I might remember it in detail." This reunion with Sadie is our first way of testing this idea, because it is the first opportunity Nanna has had for a long time to talk to someone who was there and who remembers the events. It is therefore also my first chance to see how accurate Martha's memory is. The presence of this remarkably sharp ninety-four-year-old certainly enhances my grandmother's remembering, possibly because of a consciousness that she is, by one year, the junior partner. But it works the other way around, too. One of Nanna's best friends was a lady called Nancy, who died about ten years ago. Sadie had forgotten her but has her memory thoroughly reestablished by listening to Martha's account. Martha also sets Sadie straight about the identity of the teachers in the school photo. "Do I get ten out of ten?" she jokes. Between them they

recall about twelve of their teachers, including nicknames and anecdotes in some cases, such as the improbably named woodwork master,
Mr. Woodiwiss, who used to play badminton at lunchtime with the strap-
ping Miss Ahern. Other faces they recognize but can't put a name to.
I'm not surprised that their recognition memory outstrips their recall; in
fact, it fits perfectly with the scientific evidence that recognition memory
is hardly affected by aging.

They talk for nearly two hours. Apart from a few friends of Sadie's,
neither of them has any remaining friends or family from that generation. Martha tells Sadie that she is only in the nursing home for a short
time, and talks confidently about when she will go back to her own flat.
At one point she mentions that there is a separate department in the nursing home "for the elderly," and I'm not entirely sure that it is a slip of the
tongue. Sadie mentions another name she remembers from those days:
Bertha Spanglet. There is a brief pause, and then Martha replies, "Oh yes,
yes, yes! Now when you mention the name I do remember the people . . .
She was a big girl, wasn't she?" I am watching the events of eighty years
concertina and collapse in the mind of one frail old woman. Sadie recalls
bumping into Bertha's sister recently. She says it feels like only yesterday, but in truth it must have been about twenty-five years ago. I see this
bright, articulate old woman correcting for her own distortions of time.
She knows the foibles of aging; she knows that she will underestimate
how much time has passed, and she makes allowances for it.

I come away with the feeling that I have witnessed something extraordinary. None of these jointly constructed stories have been rehearsed in
the interim, because these two people have not met for eight decades.
Adolf Hitler was just coming to power when these two last set eyes on
each other. I myself had the occasion to meet up with some old friends
recently, after a gap of twenty-seven years, which seems a mere heart-
beat in comparison. Is there any difference between meeting someone
after eight decades and being reunited with them after nearly three? I'm
not sure that our interactions worked any differently. We caught up on

the basic facts, and then started exchanging cues to memory, making stories as we went. My friends and I in our forties probably had clearer memories to work with, but we went about pooling them in essentially the same way.

Over the next couple of months I have plenty of cause to think about my interviews with Martha. I am asked to go on the radio and talk about my interviews, and they play a couple of short clips from the recordings of Martha speaking. I know that Nanna is listening to the program from her chair in the nursing home, and I phone her from Manchester straight after I have come off the air. She seems chuffed, and points out that others in the home are making a fuss of her. "I mustn't let it go to my head!" I know that our relationship has changed as a result of this project: we have seen each other more, and talked much more, and we have become closer. I suspect that she has been willing to go along with it in part because she knows her memory is failing. The stories are pretty much all she has left, and they become that much more valuable when other parts of her life are winding down. Her body is weakening—her arms are so thin, and she has diminished to the figure of a tiny bird in her chair—but she is mentally sharp, her gray eyes attentive and bright with recognition. I have had the privilege of spending time in the presence of something amazing: a mind in the act of remembering, the awesome spectacle of someone confronting their past.

I also think that Martha has tackled aspects of that past that she hasn't confronted before now. If she has any secrets, they are not about things she has done or not done—to my mind, she has lived a perfect, practically blameless life. But I suspect that there were feelings about her background, particularly about her Jewishness, that were always easier to leave unsaid. I have a strong sense that she is remembering certain things now because she *can*. As a second-generation immigrant trying to naturalize herself into English society, her Jewish identity was something that would have been almost taboo. I am observing the reversal of the same process of "willful forgetting" that has been described in writings on the diaspora by Rebecca Solnit and others. Martha's Jewish identity was something that she would have spoken about only infrequently. What is it like to be disconnected in that way from your own immediate family history? She can seem dismissive about it, sometimes almost willfully vague about basic details, such as whether her mother was from Lithuania or Russia. She tells me that her good memory has sometimes been a help to her, but at other times it has not. I have asked her many questions about memory, but at one point she asks me one in return. "Do you think childhood and growing up makes a difference to your life afterwards, when you go out into the world?" She pauses, and then says: "It must do, mustn't it?" I wonder how much this tells me about her character. I wonder what she is getting at, what knowledge remains unsaid. I suspect that she has known unkindness that she has never spoken about, and that it helped to shape the tolerant, accepting person she became.

A few weeks after the reunion with Sadie, my grandmother has a stroke that leaves her unable to speak in anything more than single syllables. The prognosis is not good for someone of her age, and complications around feeding lead to a chest infection that degenerates into double pneumonia. I drive down to see her in the hospital, knowing that it will be the last time. She is tiny under a rumpled sheet, uncomfortably pushing away at an oxygen mask, bravely trying to breathe. But she opens her eyes and smiles several times. I hold her hand, plumper and more full of life than I imagined, and I think of her story about her mother, Frieda, on the boat coming back from Lithuania on one of her trips, and how she met someone who fell in love with her hands. "She had very tiny hands," Martha once told me, "and when people used to say what small hands I had, I must have inherited hers."

She slips away that evening, with her four children at her bedside. At the funeral, amid the tears and mirthful recollections, I distribute CDs of the recordings I made with her. I am glad that we have this record, but also conscious that there are so many questions I didn't get around to asking her. I wanted to go back with her to the house on Cheshire Street, stand in the road and look up at the second-floor flat where she grew up, its London-yellow bricks with the darker brickwork arching over the windows. I wanted to walk or drive or push her in a wheelchair down Brick Lane, and ask her if she knew for sure where the bagel stall was, and what her memories of it were. I wanted to ask her whether anyone was ever really unkind to her because of who she was. I could have asked all sorts of questions that would have taught me much about memory, and much more about her. But I can't do anything now, except sit with my headphones on and interrogate the memories I already have, the clinking of teacups and the clunk of the recorder switching on and off, as Martha presses on with her endlessly renewing stories about the past.

12

A SPECIAL KIND OF TRUTH

I SET OUT to write about some science, and I ended up by telling a lot of stories. In memory, more than in any other aspect of human experience, narrative seems to be the appropriate medium. We need the science, but we also need the close attention to messy acts of meaning-making. The best memory research tries to do justice to the subjective experience as well as to the cognitive and neuroscientific mechanisms; to the stories, and what they mean to the person.

And the story continues. From the molecular level to the cultural, the science of memory has never been more vibrant. Researchers are uncovering the mysteries of the processes of protein synthesis that underlie long-term potentiation (the physical changes in synapses that lead to persistent memory traces), and the factors such as sleep that may play a role. A crucial aspect of this process, reconsolidation, provides a way of understanding the mutability of memory at a molecular level. The phenomenon of reconsolidation shows that every time a memory trace is accessed, it becomes unstable for a brief time until it can be consolidated again. That opens the door to change. In Joseph LeDoux's words, "your memory about something is only as good as your last memory about it." Catching a memory means breaking it open as well.

It is important not to get too carried away by the implications of reconsolidation. Showing that memories are mutable at the molecular level does not explain everything that we need to know about the vagaries of our autobiographical memories. Reconsolidation theory does not account for why memories change in particular ways; it only provides one possible mechanism through which that change can occur. Also, because reconsolidation operates at a different level of explanation (the molecular level), it logically dissociates from the higher-level reconstructive processes I have been writing about. You could have one without the other. A reconsolidating brain might not do much reconstruction, perhaps because it lacked the ability to integrate information from multiple cognitive and neural systems. Conversely, you could have reconstruction without reconsolidation, because the former is about the recombination of the elements of a memory, which could (logically speaking) be permanently stored.

Another big issue for the future concerns the role of the hippocampus. Some of the most exciting current work is spelling out how the hippocampus provides a spatial framework for remembering. Many questions remain, however, about the long-term role of this most essential part of the memory system. It is still not at all clear, for example, whether memories are actually stored in the hippocampus, or whether that organ's role is in binding memory features together at encoding and then providing an arena in which they can be reconstructed at retrieval. The patterns of amnesia found when there is damage to the hippocampus suggest that it cannot be the solitary nerve center of memory. At the same time, work such as the scene construction research is suggesting roles for the hippocampus that were not previously entertained.

Consider, for example, the five amnesia patients studied by Demis Hassabis and colleagues in their scene construction experiment. Four of the five patients had a problem with the future-thinking task. The fifth patient, known as P01, turned out to have profound amnesia coupled with intact future thinking. When the researchers looked more closely at P01's brain scans, they found that he retained a chunk of hippocampus on the right side. That wasn't enough to do memory, but it might have been enough to do imagination. P01's case has led Hassabis and his colleagues to speculate that the right hippocampus may have a specific role to play in imagination, while full autobiographical memory requires a functioning hippocampus on both sides.

Such a conclusion, if it is supported by future research, would confirm the intimate links between remembering and imagining. As we have seen, both of these processes depend on the ability to tell a story. So far, scientists have not paid a huge amount of attention to the importance of narrative for remembering. Redressing this balance, David Rubin has noted that narrative is a key organizational force in autobiographical memory, allowing memories to represent the passage of time and the human push toward the reaching of personal goals. Memories are told like stories, to others and to oneself. If the information does not fit with the story, as Sir Frederic Bartlett showed all those years ago, it is less likely to make it through into the memory. The neuroimaging research, together with the study of brain damage, shows that similar neural systems underpin remembering and storytelling. The main limiting factor on the emergence of autobiographical memory may be the ability to construct a narrative, known to develop in childhood later than the other components of autobiographical memory.

Memories are much more than fictional narratives, of course. Our memories are often very accurate, and only prone to serious distortion under certain conditions. To emphasize the narrative structure of memory is not to deny its potential veracity. You can make an analogy here with reportage, a dominant mode of journalism: just because it is told in the form of a story, it doesn't mean that it is not truthful. But when memory goes wrong, as in the case of some amnesias and distortions to the feeling of remembering, the stories can take over. Confabulation reminds us how the force of coherence can win over the force of correspondence, leading individuals to weave stories that fit their own reality better than they fit the reality out there.

The process works the other way around as well. Just as narrative feeds into memory, so memory feeds into narrative. Fictional writing comes alive through the inclusion of characters' memories. Would-be writers are always told to imagine what their characters think, feel and perceive, but they are not reminded often enough to give voice to their characters' memories. The work of novelists such Hilary Mantel demonstrates the power of such imagined recollections. One of the most striking things about Mantel's justly lauded novel *Wolf Hall* is the way she bestows a richly imagined past upon her protagonist, Thomas Cromwell. In one scene, Cromwell recalls an erotic encounter in a Cyprus gambling den that merges, along a link of emotion, into another sexual memory, this time in Europe, with his lover Anselma.

> *Excuse me just a moment, she had said to him; she prayed in her own language, now coaxing, now almost threatening, and she must have teased from her silver saints some flicker of grace, or perceived some deflection in their glinting rectitude, because she stood up and turned to him, saying, "I'm ready now," tugging apart the silk ties of her gown so that he could take her breasts in his hands.*

This beautiful, erotic scene is not happening in the "now" of the narrative, but in Cromwell's past. It is the fictional memory of a partly fictionalized character. Mantel presumably bases her scene on some real historical details about Cromwell's life. For the rest, she fills in the gaps with her own magic, fusing that biographical knowledge about Cromwell with sensory memories brought in from other aspects of her own experience. Other novelists, such as W. G. Sebald, create fictions that are almost memory constructions themselves: fragmentary, imagistic, fragile but striving for coherence. In Sebald's novel *Austerlitz*, for example, the protagonist's memory of his Welsh childhood is a brilliant rendering of the uncertainties and deceptions of childhood memories. In the words of the writer and psychologist Keith Oatley, Sebald's work provides us with a kind of active remembering "in which the world and selfhood are continually constructed and reconstructed—from present-day events and from not-quite-intelligible fragments of the past."

When novelists make fictional memories, they are putting together many different kinds of information, from the conceptual to the immediately experiential, and arranging them in a way that meets the needs of the present act of storytelling. (In art, in fact, you could say that the force of coherence tends to win out against the force of correspondence, whereas in science it is the other way around.) That description of fictional memory-making could be an account of how our own autobiographical memories work. As they strive to understand how the different systems of memory pull together, memory researchers could do worse than read fiction. Paying attention to how an expert novelist constructs a memory provides us with a pretty good model of our own memory systems. Accepting the narrative nature of remembering does not destroy its magic. Stories are precious, and that applies equally to our own stories of the past.

Sometimes, though, we want to be clearer that we are getting the truth rather than a story. When I read a memoir, I am always being told: This is how it was. Here is this vivid picture. Feel the weight of that vividness, its guarantee of authenticity. How could I be creating this wonderfully colorful picture if I was making it all up? But the memoirist is of course making it up. He or she is a storyteller, as we are all storytellers. I know that memory doesn't allow for that kind of faithful representation of past events. Although they are often seen in the same company, vividness does not guarantee authenticity, and Mantel is not entirely correct when, on another occasion, she complains about some of the psychological "tricks" that have been used to demonstrate the fallibility of memory. "Though my early memories are patchy," she writes, "I think they are not, or not entirely, a confabulation, and I believe this because of their overwhelming sensory power." This is an understandable mistake, but a mistake it remains. The fictions that our minds furnish us with can have *exactly* that kind of sensory power, because of the way they are concocted in our brains.

These fictions also have power because they matter. Stories and memoirs have political dimensions, and so do memories. One psychologist told me about her work researching the patterns of emotional attachment that adults display toward the key people in their lives. In order to develop expertise in measuring these patterns in her research participants, Christina enrolled in a training course for a complex but widely used attachment interview. Part of the training involved the trainees bringing in family photographs and other memorabilia of childhood. This innocuous request led Christina to a realization that would alter the course of her research, and to some extent her understanding of her own emotional life. She could not take family photographs along to the training sessions because she did not have any. They all had to be left behind when the tanks rolled into her street.

Memories work on social as well as individual levels. They can act as grievances and justifications, as weapons and as instruments of war. Christina's memories of her lost childhood home in Cyprus are part of a tapestry of similar memories that, over the years, have come to assume a tremendous political force. Palestine is another vanished land that nevertheless remains vividly present in the memories of many. A journalist friend who specializes in the area tells me that if Palestine were unremembered, a force would go out of the argument. If no one remembered the lost homes, the brutality and the terror, there would be much less to fight over.

This is too simplistic, of course, but it is a starting point. What happens to fragile human memory when it becomes politicized like this?

Does it become more vivid and particular, or more symbolic and schematic? Does it admit to uncertainty, as we have seen that it should? Robert Fisk, in his book on Lebanon, *Pity the Nation*, describes some of these acts of remembering. One elderly refugee, expelled from her Arab Palestinian village in 1948, remembers a white stone house with four upstairs and four downstairs rooms, and grapes growing up against an outside wall. Another tries to sketch a map of his stolen olive groves and becomes terribly confused, drawing map after map of half-forgotten roads. Can collective memories—constructed, mediated and negotiated as they are—deceive as much as individual memories can? If so, what does that mean for the hope of political solutions?

Perhaps we need different ways of understanding the collective memory of a society compared to those appropriate to a collection of individuals. In the civil rights movement in the United States, a shared memory of the horrors of lynching and Jim Crow laws galvanized a generation who were not even born when those events happened. How do social groups become charged with memories for events they have not actually experienced, and how do political forces act to encourage the creation of such memories? Again, the constructive view of memory allows us to understand how fragments of experience—photographs, family tales and news stories—can be smuggled into an individual's own life story. As a society, we are "remembering" all the time, falling silent on cue for Armistice Day, 9/11 and other momentous dates. A generation of Britons "remember" the anti-Vietnam Grosvenor Square protests of 1968; not all of them will have been there at the time. It is a cliché that more people claim to have memories of the Woodstock festival of 1969 than could conceivably have been present, and the same could probably be said of recent politically charged events like the 2010 tuition fee protests. They may be happening on a much grander scale, but such community acts of memory are working in the same way as my own attempts to seed memories of my father in my children's minds. Just as in that earlier, personal case, such acts come with ethical and moral responsibilities, too.

Nowhere is the political dimension of memory clearer than in its implications for our legal systems. The miscarriages of justice that have followed from too heavy a reliance on eyewitness testimony have been very thoroughly documented, and the message is getting through. The British Psychological Society recently commissioned a report, specifically aimed at those working in the criminal justice system, in which psychologists set out the facts about memory and their legal implications. In August 2011, the New Jersey Supreme Court announced sweeping changes to its treatment of eyewitness testimony, with the promise of further institutional changes to come.

The new science of memory has also raised the prospect of deliberately manipulating memory. Elizabeth Loftus has asked experimental participants whether they would take a (hypothetical) drug to erase the memory of a trauma. Eighty percent of people would not (the proportion drops when the question is framed in the context of a military scenario, where the victim has witnessed the horrors of war). When I talked to Loftus during my research for this book, she asked me the same question. I said that I too would not take the drug. I have vivid memories of certain horrible events that still upset me greatly. But I would not get rid of them. They are part of who I am; I would be less of that person if I did not have them.

My bad memories, however, are nothing like those of Colin or Peter. If I had suffered a trauma that left me unable to eat, sleep or work, I would feel differently about the preciousness of my memories. In such cases, the hypothetical scenario presented by Loftus is actually not all that far-fetched. It may not be too long before PTSD sufferers are routinely prescribed propranolol (known to block the effects of stress hormones) and other proven "memory-dampening" procedures to reduce the emotional salience of their traumatic memories as they are being re-experienced. This may not ultimately eradicate the troublesome memories, but it does seem to reduce the emotional distress they cause.

The ethics of memory manipulation, including the possibility of abuses of the technology, presents us with a huge challenge for the future. In other clinical contexts, manipulating memory would seem to be a perfectly sensible therapeutic solution. Researchers are beginning to explore the possibilities of deep-brain stimulation, where a small electrode is buried in the brain and used to stimulate relevant bits of neural circuitry, as a treatment for Alzheimer's disease. Anyone who is looking forward to a procedure that can target specific memories, however, is likely to be disappointed. One memory expert told me that he did not expect to see interventions at the level of individual memories, not in his lifetime. We simply do not know enough about how distinct events are encoded in the brain. What we do know is that the components of those mental constructions are distributed across many different cognitive and neural systems, and are thus a long way beyond our current mapping abilities.

Findings from the science of memory have also fed into wider debates about how our minds are being changed by new technologies. A media storm blew up in July 2011 when an article in the journal *Science* claimed that a reliance on Google was changing our memories. When participants knew that information was being stored by a computer, they were not as good at remembering it for themselves. They were, however, better at remembering where to go to retrieve the information afterward. The researchers saw in their findings evidence for our reliance on "transactive" memory, systems of remembering that extend beyond the limits of one individual person's brain.

Although there has been much criticism of bold claims about the effects of the Internet on our thinking, it is hard to doubt that new technologies are changing the way we outsource our memories. But then, they have always done so. The arrival of movable type printing would have transformed the world of a monk like Otgar, so heavily reliant on storing information between his own ears. When I bypass the usual process of remembering and commit an interesting website to a browser bookmark or Twitter favorite, I am simply doing what people have always done when provided with a technological prop to memory.

And it's not just technology that we turn to when our memories need help from outside. Married couples have been outsourcing memory probably for as long as humans have been getting together. I myself make no effort to remember family birthdays, because I know that Lizzie has them all down. One pair of friends does the same with London bus routes. Remembering is inherently social: just look at how it develops in childhood. It should be no surprise, therefore, that we share it around.

MEMORY, WROTE THE NOVELIST SALMAN Rushdie, has its "own special kind" of truth. "It selects, eliminates, alters, exaggerates, minimizes, glorifies and vilifies also, but in the end it creates its own reality, its heterogeneous but usually coherent version of events; and no sane human being ever trusts someone else's version more than his own." Memory may be a cheat, but it is generally speaking a beneficent one. It works tirelessly for its master.

I think that our memories *are* changing, if only because we are becoming more aware of how memory works. Although it takes us far beyond the realms of science, I am fascinated by the implications of this for the ways in which we live our lives. If our memories are constructions, incorporating plenty of genuine fact but also a healthy dose of out-and-out fiction, how does that change our relation to them? There are good reasons for clinging to the authenticity of early memories, not least because they can be so foundational for our sense of self. When you read descriptions of people's very early memories, you see that they often function as myths of creation. For Virginia Woolf, the memory of lying in her bed in the St. Ives nursery represented the moment when she became a conscious being. "If life has a base that it stands upon," she wrote, "if it is a bowl that one fills and fills and fills—then my bowl without a doubt stands upon this memory."

But embracing the constructed nature of memories can be liberating as well. I still cherish my early memories, like that of my first day at primary school. I can still hear the sound of my mother's voice, see the floating dust motes in the September-warm school hall. I just don't think it necessarily happened that way. If anything, my skepticism about memory frees me up from particular habits of remembering, which might otherwise constrain the person I believe I can be. When I talk to people about the reconstructive view of memory, I encourage them to abandon themselves to its slippery charms. We are all natural-born storytellers; we engage in acts of fiction-making every time we recount an event from our pasts. We are constantly editing and remaking our memory stories as our knowledge and emotions change. They might be fictions, but they are *our* fictions, and we should treasure them. Stories are special. Sometimes they can even be true.

I could write a whole other book on how we might use this new understanding of memory in changing our lives, but that will have to wait. Instead I want to end where I began, with the first fish I pulled out of my grandparents' lake. As we have seen throughout this book, that simple act of remembering is actually anything but simple. In having a "memory," I am drawing on visual information (the gleam of the fish, the image of the lake and its mysterious island) along with tactile details (the squidgy feel of the wet bread bait). I am using language, both to process my son's encouragement to me to reminisce, and to mediate my own thoughts about the event. I am using narrative to put it together into a story. My emotional systems are activated: I am feeling the excitement of the moment, the bigness of the world in which I am still so small. My hippocampus is laying down a spatial framework, my own internal Plan of St. Gall, to which these different elements can be assimilated. And all the while my prefrontal cortex, home of memory's search-and-retrieval processes, is busy casting its line back into the past, casting *me* back along a personal timeline in my own neural time machine.

But there is one thing that I am not doing. I am not hallucinating. I am not seven again, not really. I am two people at once: the person I am now, and the person I was then. Both individuals have a say in this

memory. Their feelings shape it, their goals structure it. It is that juxta-
position of past and present that ultimately makes it feel like a memory.
I am not reliving the experience so much as standing in a relation with it.
The memory cue (Isaac's question) comes together with the fragments of
remembered experience (and a lot of other knowledge and inference) to
create something new. Roger Shattuck, discussing Proust, put it like this:
"Like our eyes, our memories must see double; these two images then
converge in our minds into a single heightened reality." Our two eyes,
stereoscopically aligned, allow us to see space; memory allows us to "see"
time. Memories are about what happened then, but they are also about
who we are now.

BRAIN REGIONS INVOLVED IN AUTOBIOGRAPHICAL MEMORY

NOTES

Suggestions for further reading are marked in bold type.

1. CASTING A LINE

3 *memory is "a crazy woman . . ."* Austin O'Malley, *Keystones of Thought*, New York, Devin-Adair Co., 1914.

3 *some subtle internal or external connections* Donald A. Laird, "What Can You Do with Your Nose?," *The Scientific Monthly*, vol. 41, 1935.

4 *"The Memory"* A. S. Byatt, "Introduction," in **Harriet Harvey Wood and A. S. Byatt (eds.), *Memory: An Anthology*,** London, Chatto & Windus, 2008, p. xii.

4 *Without our memories, we would be lost to ourselves* Martin A. Conway, "Autobiographical Memory and Consciousness," in William P. Banks (ed.), *Encyclopedia of Consciousness*, vol. 1, Oxford, Academic Press, 2009, pp. 77–82; Joseph LeDoux, "The Self: Clues from the Brain," *Annals of the New York Academy of Sciences*, vol. 1001, 2003, pp. 295–304. For a dissenting view, see Galen Strawson, "Against Narrativity," *Ratio*, vol. 17, 2004, pp. 428–52.

5   *Metaphors of memory* Douwe Draaisma, *Metaphors of Memory: A History of Ideas about the Mind* (translated by Paul Vincent), Cambridge, Cambridge University Press, 2000.

6   *"fragile power" of memory* **Daniel L. Schacter, Searching for Memory: The Brain, the Mind, and the Past,** New York, Basic Books, 1996; John Kotre, *White Gloves: How We Create Ourselves Through Memory,* New York, The Free Press, 1995.

7   *the way a camera records them* **Daniel L. Schacter, How the Mind Forgets and Remembers: The Seven Sins of Memory,** London, Souvenir Press, 2003, p. 9.

8   *emergence of this self-understanding* Charles Fernyhough, *A Thousand Days of Wonder: A Scientist's Chronicle of His Daughter's Developing Mind,* New York, Avery, 2009.

9   *Memoir is an increasingly popular literary genre* Ian Rankin, review of *All Made Up* by Janice Galloway, *Guardian,* August 12, 2011; John Burt Foster, "Memory in the Literary Memoir," in S. Nalbantian, Paul M. Matthews and James L. McClelland (eds.), *The Memory Process: Neuroscientific and Humanistic Perspectives,* Cambridge, MA, MIT Press, 2011, pp. 297–313.

9   *"implant" memories* See note for p. 112 *the misinformation effect.*

10   *Norwegian psychologists* Svein Magnussen and Annika Melinder, "What Psychologists Know and Believe about Memory: A Survey of Practitioners," *Applied Cognitive Psychology,* vol. 26, 2012, pp. 54–60. For the full questions and correct responses, see: http://bps-research-digest.blogspot .com/2011/05/test-how-much-you-know.

10   *ordinary Americans* Daniel J. Simons and Christopher F. Chabris, "What People Believe about How Memory Works: A Representative Survey of the U.S. Population," *PLoS One,* vol. 6, 2011, e22757.

10   *Joshua Foer* Maureen Dowd, "Sexy Ruses to Stop Forgetting to Remember," *The New York Times,* March 8, 2011; Joshua Foer, *Moonwalking with Einstein: The Art and Science of Remembering Everything,* London, Allen Lane, 2011. Foer's book is focused on the "mental athletes" who employ essentially medieval techniques such as the Method of Loci to remember huge quantities of data, such as the sequences of packs of playing cards. The study of autobiographical memory arguably requires a richer conception of memory than that of "memorizing"; see e.g. Richard J. McNally, *Remembering Trauma,* Cambridge, MA, Harvard University Press, 2003; Douglas L. Hintzman, "Research Strategy in the Study of Memory: Fads,

Fallacies, and the Search for the 'Coordinates of Truth,'" *Perspectives on Psychological Science*, vol. 6, 2011, pp. 253-71.

11   *interest in memory* "Special Report: Memory," *Scientific American*, January–February 2012; "Maximising Your Memory," *Guardian*, January 14, 2012; "Put Your Memory to the Test in Our Online Experiment," *Guardian*, January 14, 2012; "The Memory Experience: A Journey of Self-discovery," BBC Radio 4, 2006.

11   *deeply held assumptions* Daniel Kahneman, *Thinking, Fast and Slow*, London, Allen Lane, 2011; David Eagleman, *Incognito: The Secret Lives of the Brain*, Edinburgh, Canongate, 2011.

11   *Galton . . . Ebbinghaus* **Douwe Draaisma, *Why Life Speeds Up as You Get Older: How Memory Shapes Our Past*** (translated by Arnold and Erica Pomerans), Cambridge, Cambridge University Press, 2004.

12   *"The War of the Ghosts"* Frederic Bartlett, *Remembering: A Study in Experimental and Social Psychology*, Cambridge, Cambridge University Press, 1950 (original work published 1932).

12   *the force of correspondence* Martin A. Conway, "Memory and the Self," *Journal of Memory and Language*, vol. 53, 2005, pp. 594–628; Bertrand Russell, *The Problems of Philosophy*, Oxford, Oxford University Press, 2001 (original work published 1912).

13   *form of long-term memory* A good introduction is **Alan Baddeley, Michael W. Eysenck and Michael C. Anderson, *Memory*,** Hove, Psychology Press, 2009.

13   *basal ganglia* Tom Hartley, Eleanor A. Maguire, Hugo J. Spiers and Neil Burgess, "The Well-worn Route and the Path Less Traveled: Distinct Neural Bases of Route Following and Wayfinding in Humans," *Neuron*, vol. 37, 2003, pp. 877–88.

13   *the engram* Richard Semon, *The Mneme*, London, George Allen & Unwin, 1921 (original work published 1904); Schacter, *Searching for Memory*; Yadin Dudai, "The Engram Revisited: On the Elusive Permanence of Memory," in Nalbantian et al. (eds.), *The Memory Process*, pp. 29–40.

13   *long-term potentiation* S. F. Cooke and T. V. P. Bliss, "Plasticity in the Human Central Nervous System," *Brain*, vol. 129, 2006, pp. 1659–73; Todd C. Sacktor, "How Does PKMζ Maintain Long-Term Memory?," *Nature Reviews Neuroscience*, vol. 12, 2011, pp. 9–15.

13   *reconsolidation* Karim Nader, Glenn E. Schafe and Joseph E. LeDoux, "Fear Memories Require Protein Synthesis in the Amygdala for Reconsolidation

After Retrieval," *Nature*, vol. 406, August 17, 2000, pp. 722–6; Karim Nader and Oliver Hardt, "A Single Standard for Memory: The Case for Reconsolidation," *Nature Reviews Neuroscience*, vol. 10, 2009, pp. 224–34.

15 *many different brain systems* Roberto Cabeza and Peggy St. Jacques, "Functional Neuroimaging of Autobiographical Memory," *Trends in Cognitive Sciences*, vol. 11, 2007, pp. 219–27. The best way to get a feel for these systems is to study a 3-D atlas. 3-D Brain is a good app for smartphones. An interactive brain map with a particular focus on memory can be found at: http://www.scientificamerican.com/article.cfm?id=memory-brain-tour-video. See diagram, p. 251.

15 *autonoetic consciousness* Endel Tulving, "Memory and Consciousness," *Canadian Psychology*, vol. 26, 1985, pp. 1–12.

16 *participants' individual stories* Hintzman, "Research Strategy in the Study of Memory."

2. GETTING LOST

22 *stimuli that you are exposed to* R. S. Nickerson and M. J. Adams, "Long-term Memory for a Common Object," *Cognitive Psychology*, vol. 11, 1979, pp. 287–307.

24 *reminiscence bump* David C. Rubin and Matthew D. Schulkind, "The Distribution of Autobiographical Memories across the Lifespan," *Memory & Cognition*, vol. 25, 1997, pp. 859–66.

26 *lost has two different meanings* Rebecca Solnit, *A Field Guide to Getting Lost*, Edinburgh, Canongate, 2006, p. 22.

27 *"There's an art . . . to attending to weather . . ."* Ibid., p. 10.

28 *people do indeed walk in circles* Jan L. Souman, Ilja Frissen, Manish N. Sreenivas and Marc O. Ernst, "Walking Straight into Circles," *Current Biology*, vol. 19, 2009, pp. 1538–42.

29 *place cells . . . head-direction cells* John O'Keefe and Lynn Nadel, *The Hippocampus as a Cognitive Map*, Oxford, Clarendon, 1978; Jeffrey S. Taube, "Head Direction Cells and the Neurophysiological Basis for a Sense of Direction," *Progress in Neurobiology*, vol. 55, 1998, pp. 225–56.

30 *grid cells* Torkel Hafting, Marianne Fyhn, Sturla Molden, May-Britt Moser and Edvard I. Moser, "Microstructure of a Spatial Map in the Entorhinal Cortex," *Nature*, vol. 436, August 11, 2005, pp. 801–6.

30   *similar hexagonal grids* Christian F. Doeller, Caswell Barry and Neil Burgess, "Evidence for Grid Cells in a Human Memory Network," *Nature*, vol. 463, February 4, 2010, pp. 657–61.

31   *as the crow flies* Moheb Costandi, "Human Brain Maps Flip during Spatial Navigation," Action Potential, November 18, 2011, http://blogs.nature.com/actionpotential/2011/11/human_brain_maps_flip_during_s.html.

31   *hippocampal theta* György Buzsáki, "Neurons and Navigation," *Nature*, vol. 436, August 11, 2005, pp. 781–2.

35   *the blue hotel* Charles Fernyhough, *The Auctioneer*, London, Fourth Estate, 1999.

3. THE SCENT MUSEUM

42   *In Marcel Proust's masterpiece* Marcel Proust, *In Search of Lost Time* (translated by C. K. Scott Moncrieff and Terence Kilmartin; revised by D. J. Enright), New York, The Modern Library, 1998 (original work published 1913), pp. 60–64; J. Stephan Jellinek, "Proust Remembered: Has Proust's Account of Odor-cued Autobiographical Memory Recall Really Been Investigated?," *Chemical Senses*, vol. 29, 2004, pp. 455–8; Douwe Draaisma, *Why Life Speeds Up as You Get Older: How Memory Shapes Our Past* (translated by Arnold and Erica Pomerans), Cambridge, Cambridge University Press, 2004.

44   *the reconstructive nature of memory* Jonah Lehrer, *Proust Was a Neuroscientist*, Edinburgh, Canongate, 2011.

45   *odd neurological properties* David Bainbridge, *Beyond the Zonules of Zinn: A Fantastic Journey through Your Brain*, Cambridge, MA, Harvard University Press, 2008.

45   *as much as he was tasting it* Dana M. Small, Johannes C. Gerber, Y. Erica Mak and Thomas Hummel, "Differential Neural Responses Evoked by Orthonasal versus Retronasal Odorant Perception in Humans," *Neuron*, vol. 47, 2005, pp. 593–605.

45   *more visual writer* Avery Gilbert, *What the Nose Knows: The Science of Scent in Everyday Life*, New York, Crown, 2008.

46   *smell seems to shift the reminiscence bump* Simon Chu and John Joseph Downes, "Long Live Proust: The Odour-cued Autobiographical Memory Bump," *Cognition*, vol. 75, 2000, B41–B50; Amanda N. Miles and Dorthe Berntsen, "Odour-induced Mental Time Travel into the Past and Future: Do

Odour Cues Retain a Unique Link to Our Distant Past?," *Memory*, vol. 19, 2011, pp. 930–40; Maria Larsson and Johan Willander, "Autobiographical Odor Memory," *International Symposium on Olfaction and Taste: Annals of the New York Academy of Sciences*, vol. 1170, 2009, pp. 318–23.

46 *a certain keen-smelling mammal* Richard Holmes, "A Meander through Memory and Forgetting," in Harriet Harvey Wood and A. S. Byatt (eds.), *Memory: An Anthology*, London, Chatto & Windus, 2008.

47 *popcorn, freshly cut grass* Rachel S. Herz, "A Naturalistic Analysis of Autobiographical Memories Triggered by Olfactory Visual and Auditory Stimuli," *Chemical Senses*, vol. 29, 2004, pp. 217–24; Rachel S. Herz, James Eliassen, Sophia Beland and Timothy Souza, "Neuroimaging Evidence for the Emotional Potency of Odor-evoked Memory," *Neuropsychologia*, vol. 42, 2004, pp. 371–8.

47 *there is still some debate* Catherine de Lange, "The Unsung Sense: How Smell Rules Your Life," *New Scientist*, September 19, 2011; Marieke B. J. Toffolo, Monique A. M. Smeets and Marcel A. van den Hout, "Proust Revisited: Odours as Triggers of Aversive Memories," *Cognition and Emotion*, vol. 26, 2012, pp. 83–92.

47 *for the imagining of future events* Amanda N. Miles and Dorthe Berntsen, "Odour-induced Mental Time Travel into the Past and Future: Do Odour Cues Retain a Unique Link to Our Distant Past?," *Memory*, vol. 19, 2011, pp. 930–40.

48 *Proust did not break new ground* Gilbert, *What the Nose Knows*; J. Bogousslavsky and O. Walusinski, "Marcel Proust and Paul Sollier: The Involuntary Memory Connection," *Schweizer Archiv für Neurologie und Psychiatrie*, vol. 160, 2009, pp. 130–36; Jean Delacour, "Proust's Contribution to the Psychology of Memory: The Reminiscences from the Standpoint of Cognitive Science," *Theory & Psychology*, vol. 11, 2001, pp. 255–71.

49 *scent museum* Andy Warhol, *The Philosophy of Andy Warhol*, London, Penguin, 2007 (original work published 1975), p. 151.

50 *a bunch of sage brush* Donald A. Laird, "What Can You Do with Your Nose?," *The Scientific Monthly*, vol. 41, 1935, p. 128.

50 *those studying for exams* Rachel Herz, *The Scent of Desire: Discovering Our Enigmatic Sense of Smell*, New York, Harper Perennial, 2008.

50 *Marcel exploits his own forgetfulness* Roger Shattuck, *Proust's Way: A Field Guide to* In Search of Lost Time, London, Allen Lane, 2000. See also note for p. 186 *suppress other memories*.

50 *involuntary memory* John H. Mace, "Involuntary Autobiographical Memories Are Highly Dependent on Abstract Cuing: The Proustian View Is Incorrect," *Applied Cognitive Psychology*, vol. 18, 2004, pp. 893–9.

51 *more likely to be triggered by abstract cues* Dorthe Berntsen and Nicoline Marie Hall, "The Episodic Nature of Involuntary Autobiographical Memories," *Memory & Cognition*, vol. 32, 2004, pp. 789–803.

51 *smell memories are susceptible to interference* Trygg Engen and Bruce M. Ross, "Long-term Memory of Odors with and without Verbal Descriptions," *Journal of Experimental Psychology*, vol. 100, 1973, pp. 221–7; Theresa L. White, "Olfactory Memory: The Long and the Short of It," *Chemical Senses*, vol. 23, 1998, pp. 433–41.

51 *four hundred different olfactory genes* Catherine de Lange, "The Unsung Sense: How Smell Rules Your Life," *New Scientist*, September 19, 2011.

52 *early smell memory* Yaara Yeshurun, Hadas Lapid, Yadin Dudai and Noam Sobel, "The Privileged Brain Representation of First Olfactory Associations," *Current Biology*, vol. 19, 2009, pp. 1869–74.

53 *"they cover their own tracks . . ."* Gilbert, *What the Nose Knows*, p. 201.

54 *bits of music* Jaclyn Hennessey Ford, Donna Rose Addis and Kelly S. Giovanello, "Differential Neural Activity during Search of Speci and General Autobiographical Memories Elicited by Musical Cues," *Neuropsychologia*, vol. 49, 2011, pp. 2514–26.

55 *when a smell, for example, is paired with a picture* Jay A. Gottfried, Adam P. R. Smith, Michael D. Rugg and Raymond J. Dolan, "Remembrance of Odors Past: Human Olfactory Cortex in Cross-modal Recognition Memory," *Neuron*, vol. 42, 2004, pp. 687–95.

4. THE SUNNY NEVER-NEVER

61 *observer memories . . . field memories* The reconstructive view of observer memories is supported by findings that their likelihood of occurring depends on what question is being asked. This study showed that field (as opposed to observer) memories were more likely to occur when participants were asked to focus on emotions rather than objective details: Georgia Nigro and Ulric Neisser, "Point of View in Personal Memories," *Cognitive Psychology*, vol. 15, 1983, pp. 467–82. Distinct neural systems have been proposed to mediate field and observer memories: Eric Eich, Andrea L. Nelson, M. Adil Leghari and Todd C. Handy, "Neural Systems

Mediating Field and Observer Memories," *Neuropsychologia*, vol. 47, 2009, pp. 2239–51.

61 *bland, third-person memories acted as screens* Sigmund Freud, "Screen Memories," in J. Strachey (ed. and trans.), *The Standard Edition of the Complete Psychological Works of Sigmund Freud*, vol. 3, London, The Hogarth Press, 1975 (original work published 1899).

62 *"a dog that lies down where it pleases"* Cees Nooteboom, *Rituals* (translated by Adrienne Dixon), London, Penguin, 1985.

62 *early insults to our egos* Willem A. Wagenaar, "Remembering My Worst Sins: How Autobiographical Memory Serves the Updating of the Conceptual Self," in M. A. Conway, D. C. Rubin, H. Spinnler and W. A. Wagenaar (eds.), *Theoretical Perspectives on Autobiographical Memory*, Dordrecht, Springer, 1992; Douwe Draaisma, *Why Life Speeds Up as You Get Older: How Memory Shapes Our Past* (translated by Arnold and Erica Pomerans), Cambridge, Cambridge University Press, 2004.

62 *"the remarkable amnesia of childhood"* Sigmund Freud, in J. Strachey (ed. and trans.), *The Standard Edition of the Complete Psychological Works of Sigmund Freud*, vol. 15, London, Penguin, 1963, pp. 199–200.

62 *earliest memories* Darryl Bruce, L. Amber Wilcox-O'Hearn, John A. Robinson, Kimberly Phillips-Grant, Lori Francis and Marilyn C. Smith, "Fragment Memories Mark the End of Childhood Amnesia," *Memory & Cognition*, vol. 33, 2005, pp. 567–76; M. J. Eacott and R. A. Crawley, "The Offset of Childhood Amnesia: Memory for Events That Occurred before Age 3," *Journal of Experimental Psychology: General*, vol. 127, 1998, pp. 22–33.

63 *operant conditioning* Carolyn Rovee-Collier and Peter Gerhardstein, "The Development of Infant Memory," in Nelson Cowan (ed.), *The Development of Memory in Childhood*, Hove, Psychology Press, 1997.

64 *fetuses can learn while still in the womb* P. G. Hepper, "Fetal Memory: Does It Exist? What Does It Do?," *Acta Paediatrica Supplement*, vol. 416, 1996, pp. 16–20.

64 *remembering and forgetting in babyhood* Carolyn Rovee-Collier and Harlene Hayne, "Memory in Infancy and Early Childhood," in Endel Tulving and Fergus I. M. Craik (eds.), *The Oxford Handbook of Memory*, Oxford, Oxford University Press, 2000; Patricia J. Bauer, "Constructing a Past in Infancy: A Neuro-Developmental Account," *Trends in Cognitive Sciences*,

vol. 10, 2006, pp. 175–81; Kirsten Weir, "Infant Recall: The Birth of Memory," *New Scientist*, May 6, 2011.

64 *As soon as you can use words to describe your experience* Catriona M. Morrison and Martin A. Conway, "First Words and First Memories," *Cognition*, vol. 116, 2010, pp. 23–32; Weir, "Infant Recall."

65 *always at home in our pasts* Vladimir Nabokov, *Speak, Memory: An Autobiography Revisited*, London, Penguin, 2000 (original work published 1967).

65 *Robert Fripp recalled as his first memory* Interview in *Guitar Player*, January 1986, p. 97.

66 *Edith Wharton, Anthony Powell* Edith Wharton, *A Backward Glance: An Autobiography*, New York, Touchstone, 1998 (original work published 1933); Anthony Powell, *Infants of the Spring*, London, Heinemann, 1976; Mark L. Howe and Mary L. Courage, "The Emergence and Early Development of Autobiographical Memory," *Psychological Review*, vol. 104, 1997, pp. 499–523.

66 *parrotings of things their parents said* Catherine A. Haden, Peter A. Ornstein, Carol O. Eckerman and Sharon M. Didow, "Mother-Child Conversational Interactions as Events Unfold: Linkages to Subsequent Remembering," *Child Development*, vol. 72, 2001, pp. 1016–31; Katherine Nelson and Robyn Fivush, "The Emergence of Autobiographical Memory: A Social Cultural Developmental Theory," *Psychological Review*, vol. 111, 2004, pp. 486–511.

67 *"Some of the things which one seems to remember . . ."* Leonard Woolf, *Sowing: An Autobiography of the Years 1880–1904*, London, The Hogarth Press, 1960.

67 *relatively weak capacities to monitor the sources* D. Stephen Lindsay, Marcia K. Johnson and Paul Kwon, "Developmental Changes in Memory Source Monitoring," *Journal of Experimental Child Psychology*, vol. 52, 1991, pp. 297–318.

67 *digital representations of the events* Charles Fernyhough, "Images of Childhood," *Financial Times*, May 23, 2009.

67 *"mental time travel"* Endel Tulving, *Elements of Episodic Memory*, Oxford, Clarendon, 1983; Thomas Suddendorf and Michael C. Corballis, "The Evolution of Foresight: What Is Mental Time Travel, and Is It Unique to Humans?," *Behavioral and Brain Sciences*, vol. 30, 2007, pp. 299–351.

68 *prefrontal cortex* Aleksandr R. Luria, *The Working Brain: An Introduction to Neuropsychology* (translated by B. Haigh), New York, Basic Books, 1973.

69 *the boundary of childhood amnesia seems to shift* Karen Tustin and Harlene Hayne, "Defining the Boundary: Age-related Changes in Childhood Amnesia," *Developmental Psychology*, vol. 46, 2010, pp. 1049–61; Carole Peterson, Valerie V. Grant and Lesley D. Boland, "Childhood Amnesia in Children and Adolescents: Their Earliest Memories," *Memory*, vol. 13, 2005, pp. 622–37; Carole Peterson, Kelly L. Warren and Megan M. Short, "Infantile Amnesia across the Years: A 2-Year Follow-up of Children's Earliest Memories," *Child Development*, vol. 82, 2011, pp. 1092–105. An earlier version of this section appeared in the blog post "The Shifting Boundary of Childhood Amnesia," December 2, 2010, http://www.psychologytoday.com/blog/the-child-in-time/201012/the-shifting-boundary-childhood-amnesia.

70 *study of Michael's memory* Aletha Solter, "A 2-Year-Old Child's Memory of Hospitalization during Early Infancy," *Infant and Child Development*, vol. 17, 2008, pp. 593–605. An earlier version of this section appeared in the blog post "Michael's Memory," January 24, 2009, http://pieceslight.blogspot.co.uk/2009/01/michael-memory.html.

71 *interesting bubble-making machine* Gwynn Morris and Lynne Baker-Ward, "Fragile But Real: Children's Capacity to Use Newly Acquired Words to Convey Preverbal Memories," *Child Development*, vol. 78, 2007, pp. 448–58; although see Gabrielle Simcock and Harlene Hayne, "Breaking the Barrier? Children Fail to Translate Their Preverbal Memories into Language," *Psychological Science*, vol. 13, 2002, pp. 225–31.

71 *cast a line back into babyhood* Nancy Angrist Myers, Rachel Keen Clifton and Marsha G. Clarkson, "When They Were Very Young: Almost-Threes Remember Two Years Ago," *Infant Behavior and Development*, vol. 10, 1987, pp. 123–32.

71 *a chance to rehearse, consolidate and talk about memory* Bauer, "Constructing a Past in Infancy."

72 *she could remember nothing of her infancy* Charles Fernyhough, *A Thousand Days of Wonder: A Scientist's Chronicle of His Daughter's Developing Mind*, New York, Avery, 2009.

72 *medial temporal and frontal lobes* Bauer, "Constructing a Past in Infancy."

74 ". . . *series of spaced flashes* . . ." Nabokov, *Speak, Memory*, p. 18.

74 "... *the brightness of light* ..." Georgia O'Keeffe, *Georgia O'Keeffe*, New York, Penguin, 1977, p. 1.

74 "*fragment memories*" Bruce et al., "Fragment Memories Mark the End of Childhood Amnesia."

75 *Two of the most famous accounts* Virginia Woolf, "A Sketch of the Past," in *Moments of Being* (2nd edition), Orlando, FL, Harcourt Brace & Company, 1985, p. 64. Woolf's honesty about what is "convenient artistically" illustrates the tension between the forces of coherence and correspondence in memory; see note for p. 12 *the force of correspondence* and Charles Fernyhough, "The Story of the Self," *Guardian*, January 14, 2012.

76 "*isolated spots or patches* ..." W. H. Hudson, *Far Away and Long Ago: A History of My Early Life*, London, J. M. Dent & Sons, 1918, p. 2.

5. WALKING AT GOLDHANGER

80 *levels of processing* Fergus I. M. Craik and Endel Tulving, "Depth of Processing and the Retention of Words in Episodic Memory," *Journal of Experimental Psychology: General*, vol. 104, 1975, pp. 268–94; Fergus I. M. Craik and Robert S. Lockhart, "Levels of Processing: A Framework for Memory Research," *Journal of Verbal Learning and Verbal Behavior*, vol. 11, 1972, pp. 671–84; Bradford H. Challis, Boris M. Velichkovsky and Fergus I. M. Craik, "Levels-of-Processing Effects on a Variety of Memory Tasks: New Findings and Theoretical Implications," *Consciousness and Cognition*, vol. 5, 1996, pp. 142–64.

81 "*verbatim effect*" Jacqueline Strunk Sachs, "Recognition Memory for Syntactic and Semantic Aspects of Connected Discourse," *Perception & Psychophysics*, vol. 2, 1967, pp. 437–42; J. Poppenk, G. Walia, A. R. McIntosh, M. F. Joanisse, D. Klein and S. Köhler, "Why Is the Meaning of a Sentence Better Remembered than Its Form? An fMRI Study on the Role of Novelty-encoding Processes," *Hippocampus*, vol. 18, 2008, pp. 909–18.

81 *more likely to commit new information to memory* Endel Tulving and Neal Kroll, "Novelty Assessment in the Brain and Long-term Memory Encoding," *Psychonomic Bulletin & Review*, vol. 2, 1995, pp. 387–90.

82 *The hippocampus also plays a crucial role* Katharina Henke, Bruno Weber, Stefan Kneifel, Heinz Gregor Wieser and Alfred Buck, "Human Hippocampus Associates Information in Memory," *PNAS*, vol. 96, 1999, pp. 5884–9.

85   *This is not a judgment based on the absence of familiarity, or even a posi-*
      tive feeling of unfamiliarity Jane Plailly, Barbara Tillmann and Jean-Pierre
      Royet, "The Feeling of Familiarity of Music and Odors: The Same Neural
      Signature?," *Cerebral Cortex*, vol. 17, 2007, pp. 2650–58.

85   *pigeons learned to discriminate* William Vaughan, Jr., and Sharon L.
      Greene, "Pigeon Visual Memory Capacity," *Journal of Experimental Psy-
      chology: Animal Behavior Processes*, vol. 10, 1984, pp. 256–71.

87   *two neurally distinct processes* Michael D. Rugg and Tim Curran, "Event-
      related Potentials and Recognition Memory," *Trends in Cognitive Sciences*,
      vol. 11, 2007, pp. 251–7.

87   *You can recognize a place because it is familiar* Daniela Montaldi and Andrew
      R. Mayes, "The Role of Recollection and Familiarity in the Functional Dif-
      ferentiation of the Medial Temporal Lobes," *Hippocampus*, vol. 20, 2010,
      pp. 1291–314; Rachel A. Diana, Andrew P. Yonelinas and Charan Ranga-
      nath, "Imaging Recollection and Familiarity in the Medial Temporal Lobe:
      A Three-component Model," *Trends in Cognitive Sciences*, vol. 11, 2007,
      pp. 379–86. Note that, although the perirhinal cortex has been clearly linked
      to object familiarity, its role in scene familiarity is less certain at present.

91   *deep-sea divers* D. R. Godden and A. D. Baddeley, "Context-dependent
      Memory in Two Natural Environments: On Land and Underwater," *Brit-
      ish Journal of Psychology*, vol. 66, 1975, pp. 325–31.

91   *cues that are around at the moment of encoding* Endel Tulving and
      Donald M. Thomson, "Encoding Specificity and Retrieval Processes in
      Episodic Memory," *Psychological Review*, vol. 80, 1973, pp. 352–73.

6. NEGOTIATING THE PAST

95   *"worthless straws"* Samuel Taylor Coleridge, "Defects of Wordsworth's
      Poetry," in James Engell and W. Jackson Bate (eds.), *The Collected Works of
      Samuel Taylor Coleridge*, vol. 7, Princeton, NJ, Princeton University Press,
      1983 (original work published 1817), p. 139.

96   *structured systems of remembering* Darryl Bruce, L. Amber Wilcox-
      O'Hearn, John A. Robinson, Kimberly Phillips-Grant, Lori Francis and
      Marilyn C. Smith, "Fragment Memories Mark the End of Childhood
      Amnesia," *Memory & Cognition*, vol. 33, 2005.

97   *first references to past events* Katherine Nelson and Robyn Fivush, "Social-
      ization of Memory," in E. Tulving and F. I. M. Craik (eds.), *Oxford Hand-
      book of Memory*, Oxford, Oxford University Press, 2000.

97 *effects of parental elaboration* Elaine Reese, Catherine A. Haden and Robyn Fivush, "Mother-Child Conversations about the Past: Relationships of Style and Memory over Time," *Cognitive Development*, vol. 8, 1993, pp. 403–30; Catherine A. Haden, Rachel A. Haine and Robyn Fivush, "Developing Narrative Structure in Parent-Child Reminiscing across the Preschool Years," *Developmental Psychology*, vol. 33, 1997, pp. 295–307; Fiona Jack, Shelley MacDonald, Elaine Reese and Harlene Hayne, "Maternal Reminiscing Style during Early Childhood Predicts the Age of Adolescents' Earliest Memories," *Child Development*, vol. 80, 2009, pp. 496–505. There are pronounced cultural differences in parent-child conversations about the past; see for example Qi Wang, "Relations of Maternal Style and Child Self-concept to Autobiographical Memories in Chinese, Chinese Immigrant, and European American 3-Year-Olds," *Child Development*, vol. 77, 2006, pp. 1794–809.

98 *something particularly pure about early fragment memories* Virginia Woolf, "A Sketch of the Past," in *Moments of Being* (2nd edition), Orlando, FL, Harcourt Brace & Company, 1985, p. 67.

99 *High-quality memories* Patricia J. Bauer, "Constructing a Past in Infancy: A Neuro-Developmental Account," *Trends in Cognitive Sciences*, vol. 10, 2006.

100 *romanticizing past relationships* Erik Hesse, "The Adult Attachment Interview: Protocol, Method of Analysis, and Empirical Studies," in J. Cassidy and P. R. Shaver (eds.), *Handbook of Attachment: Theory, Research, and Clinical Applications* (2nd edition), London, The Guilford Press, 2008, pp. 552–98.

101 *young children are "dualists"* Paul Bloom, *Descartes' Baby: How Child Development Explains What Makes Us Human*, London, William Heinemann, 2004.

101 *children who understand biological death* Jesse M. Bering and David F. Bjorklund, "The Natural Emergence of Reasoning about the Afterlife as a Developmental Regularity," *Developmental Psychology*, vol. 40, 2004, pp. 217–33; Paul L. Harris and Marta Giménez, "Children's Acceptance of Conflicting Testimony: The Case of Death," *Journal of Cognition and Culture*, vol. 5, 2005, pp. 143–64.

102 *first narrative constructions* Stephen J. Ceci and Maggie Bruck, "Suggestibility of the Child Witness: A Historical Review and Synthesis," *Psychological Bulletin*, vol. 113, 1993, pp. 403–39; Charles Fernyhough, "Images of Childhood," *Financial Times*, May 23, 2009. An earlier version of this

section appeared as: Charles Fernyhough, "Grandad: Back from the Dead," *Guardian*, October 3, 2009.

105 *"A happy love is a single story . . ."* Rebecca Solnit, *A Field Guide to Getting Lost*, Edinburgh, Canongate, 2006, p. 135.

105 *"siblings battle . . ."* Dorothy Rowe, *My Dearest Enemy, My Dangerous Friend: Making and Breaking Sibling Bonds*, London, Routledge, 2007, p. xii.

108 *"Her memories are so twisted . . ."* Terri Apter, *The Sister Knot*, New York, Norton, 2007, p. 225.

109 *who "owns" a memory* Jonathan Margolis, "Confessions of a Celebrity Biographer," *Guardian*, August 4, 2010; Jonathan Heawood, "A Privacy Law Must Not Muzzle Our Memories," *Guardian*, December 28, 2011.

109 *disputes over memory ownership* Mercedes Sheen, Simon Kemp and David Rubin, "Twins Dispute Memory Ownership: A New False Memory Phenomenon," *Memory & Cognition*, vol. 29, 2001, pp. 779–88; M. Sheen, S. Kemp and D. C. Rubin, "Disputes over Memory Ownership: What Memories Are Disputed?," *Genes, Brain and Behavior*, vol. 5 (suppl. 1), 2006, pp. 9–13.

110 *"Well, that actually happened to me if you don't mind . . ."* Ibid., p. 9.

111 *a perpetually cross person* A. S. Byatt, "Introduction," in Harriet Harvey Wood and A. S. Byatt (eds.), *Memory: An Anthology*, London, Chatto & Windus, 2008, p. xiii.

111 *a ride in a hot-air balloon* Kimberley A. Wade, Maryanne Garry, J. Don Read and D. Stephen Lindsay, "A Picture Is Worth a Thousand Lies: Using False Photographs to Create False Childhood Memories," *Psychonomic Bulletin & Review*, vol. 9, 2002, pp. 597–603.

112 *the misinformation effect* Yoko Okado and Craig E. L. Stark, "Neural Activity during Encoding Predicts False Memories Created by Misinformation," *Learning & Memory*, vol. 12, 2005, pp. 3–11; Elizabeth F. Loftus, "Planting Misinformation in the Human Mind: A 30-year Investigation of the Malleability of Memory," *Learning & Memory*, vol. 12, 2005, pp. 361–6.

112 *People have recalled nonexistent objects such as broken glass* Ibid., p. 361.

113 *when our memories are shaped by other people* Suparna Rajaram, "Collaboration Both Hurts and Helps Memory: A Cognitive Perspective," *Current Directions in Psychological Science*, vol. 20, pp. 76–81; Henry L. Roediger III and Kathleen B. McDermott, "Remember When?," *Science*, vol. 333, July 1, 2011, pp. 47–8; Micah Edelson, Tali Sharot, Raymond J. Dolan and Yadin Dudai, "Following the Crowd: Brain Substrates of Long-term Memory Conformity," *Science*, vol. 333, July 1, 2011, pp. 108–11.

115 *"nonbelieved memories"* Giuliana Mazzoni, Alan Scoboria and Lucy Harvey, "Nonbelieved Memories," *Psychological Science*, vol. 21, 2010, pp. 1334–40. An earlier version of this section appeared in the blog post "Remembering Events That Never Happened," February 17, 2011, http://www.psychologytoday.com/blog/the-child-in-time/201102/remembering-events-never-happened.

7. THE PLAN OF WHAT MIGHT BE

121 *this movement of the mind toward God* Otgar is a fictional ninth-century monk. In creating his story I have drawn on numerous texts including the following: Mary Carruthers, *The Craft of Thought: Meditation, Rhetoric, and the Making of Images, 400–1200*, Cambridge, Cambridge University Press, 1998; Mary Carruthers and Jan M. Ziolkowski, *The Medieval Craft of Memory: An Anthology of Texts and Pictures*, Philadelphia, University of Pennsylvania Press, 2002; Mary Carruthers, *The Book of Memory: A Study of Memory in Medieval Culture* (2nd edition), Cambridge, Cambridge University Press, 1990.

122 *The Plan of St. Gall* Carruthers, *The Craft of Thought*. The Plan of St. Gall can be viewed at http://www.stgallplan.org/en/index.html.

122 *Memoria, in Carruthers's analysis* Carruthers, *The Craft of Thought*; **Yadin Dudai and Mary Carruthers, "The Janus Face of Mnemosyne,"** *Nature*, vol. 434, 2005, p. 567.

122 *"the craft of making thoughts about God"* Carruthers, *The Craft of Thought*, p. 2.

122 *"... run freely through open space ..."* Carruthers and Ziolkowski, *The Medieval Craft of Memory*, pp. 2–3.

123 *"... a universal thinking machine ..."* Carruthers, *The Craft of Thought*, p. 4.

124 *"Funes the Memorious"* Jorge Luis Borges, "Funes, His Memory," in *Collected Fictions*, London, Allen Lane, 1998, pp. 131–7.

124 *hyperthymestic syndrome* Elizabeth S. Parker, Larry Cahill and James L. McGaugh, "A Case of Unusual Autobiographical Remembering," *Neurocase*, vol. 12, 2006, pp. 35–49.

125 *Presented with a list of words on a similar theme* Daniel L. Schacter, Joan Y. Chiao and Jason P. Mitchell, "The Seven Sins of Memory: Implications for Self," *Annals of the New York Academy of Sciences*, vol. 1001, 2003, pp. 226–39; Henry L. Roediger III and Kathleen B. McDermott, "Creating

False Memories: Remembering Words Not Presented in Lists," *Journal of Experimental Psychology: Learning, Memory, and Cognition*, vol. 21, 1995, pp. 803–14; **Daniel L. Schacter and Donna Rose Addis, "The Ghosts of Past and Future,"** *Nature*, vol. 445, 2007, p. 27.

125 *As Xenophon put it* Carruthers, *The Craft of Thought*, p. 30.

126 *". . . adaptive, constructive processes . . ."* Schacter and Addis, "The Ghosts of Past and Future," p. 27.

126 *sins of memory* Daniel L. Schacter, *How the Mind Forgets and Remembers: The Seven Sins of Memory*, London, Souvenir Press, 2003.

126 *clues to why it might have evolved* Mark L. Howe, "The Adaptive Nature of Memory and Its Illusions," *Current Directions in Psychological Science*, vol. 20, 2011, pp. 312–15.

127 *function of episodic memory* Martin A. Conway, "Episodic Memories," *Neuropsychologia*, vol. 47, 2009, pp. 2305–13.

127 *future scenarios* D. H. Ingvar, "'Hyperfrontal' Distribution of the Cerebral Grey Matter Flow in Resting Wakefulness: On the Functional Anatomy of the Conscious State," *Acta Neurologica Scandinavica*, vol. 60, 1979, pp. 12–25; D. H. Ingvar, "Memory of the Future: An Essay on the Temporal Organization of Conscious Awareness," *Human Neurobiology*, vol. 4, pp. 127–36.

127 *cognitive and neural systems* Daniel L. Schacter, Donna Rose Addis and Randy L. Buckner, "Remembering the Past to Imagine the Future: The Prospective Brain," *Nature Reviews Neuroscience*, vol. 8, pp. 657–61; Stefania de Vito, "Images of the Future, Drawn from the Past," *The Psychologist*, vol. 23, 2010, pp. 570–72; Daniel L. Schacter, Donna Rose Addis and Randy L. Buckner, "Episodic Simulation of Future Events," *Annals of the New York Academy of Sciences*, vol. 1124, 2008, pp. 39–60.

127 *core memory system* Randy L. Buckner and Daniel C. Carroll, "Self-projection and the Brain," *Trends in Cognitive Sciences*, vol. 11, 2007, pp. 49–57. Note the overlap between the core memory system and the "default mode network," shown to activate when participants are engaged in passive, non-task-directed thinking.

127 *future scenarios are remembered* Victoria C. Martin, Daniel L. Schacter, Michael C. Corballis and Donna Rose Addis, "A Role for the Hippocampus in Encoding Simulations of Future Events," *PNAS*, vol. 108, 2011, pp. 13858–63.

128   *bias toward positive future scenarios* Tali Sharot, *The Optimism Bias: Why We're Wired to Look on the Bright Side*, London, Robinson, 2012.

128   *emotional content of future simulations* Karl K. Szpunar, Donna Rose Addis and Daniel L. Schacter, "Memory for Emotional Simulations: Remembering a Rosy Future," *Psychological Science*, vol. 23, 2012, pp. 24–9.

128   *difficulty in predicting the future* Demis Hassabis, Dharshan Kumaran, Seralynne D. Vann and Eleanor A. Maguire, "Patients with Hippocampal Amnesia Cannot Imagine New Experiences," *PNAS*, vol. 104, 2007, pp. 1726–31. For further discussion of patient P01, see chapter 12. For contrary findings on amnesia and future-related thinking, see Larry R. Squire, Anna S. van der Horst, Susan G. R. McDuff, Jennifer C. Frascino, Ramona O. Hopkins and Kristin N. Mauldin, "Role of the Hippocampus in Remembering the Past and Imagining the Future," *PNAS*, vol. 107, 2010, pp. 19044–8. For details on the role played by other cortical regions (such as the parietal lobe) in memory construction, see Marian E. Berryhill, Lisa Phuong, Lauren Picasso, Roberto Cabeza and Ingrid R. Olson, "Parietal Lobe and Episodic Memory: Bilateral Damage Causes Impaired Free Recall of Autobiographical Memory," *Journal of Neuroscience*, vol. 27, 2007, pp. 14415–23.

129   *ten most important scientific breakthroughs of 2007* The News Staff, "Breakthrough of the Year: The Runners-Up," *Science*, vol. 318, 2007, pp. 1844–9.

130   *scene construction* Demis Hassabis and Eleanor A. Maguire, "The Construction System of the Brain," *Philosophical Transactions of the Royal Society B*, vol. 364, 2009, pp. 1263–71; **Demis Hassabis and Eleanor A. Maguire, "Deconstructing Episodic Memory with Construction,"** *Trends in Cognitive Sciences*, vol. 11, 2007, pp. 299–306.

130   *hippocampus "providing the spatial backdrop . . ."* Roger Highfield, "Mapping Memories: Eleanor Maguire and Brain Imaging," http://www.wellcome.ac.uk/About-us/75th-anniversary/WTVM052023.htm.

130   *good at processing spatial information* Joshua Foer, *Moonwalking with Einstein: The Art and Science of Remembering Everything*, London, Allen Lane, 2011.

130   *"memory palaces"* Frances A. Yates, *The Art of Memory*, London, Pimlico, 1992 (original work published 1966). While recognizing her immense contribution to the field, Mary Carruthers (*The Craft of Thought*, p. 9) is critical of Yates for her failure to acknowledge sufficiently the creative, combinatorial nature of the medieval art of memory. See note for p. 10 *Joshua Foer*.

131 *memory isn't . . . time sensitive* Hassabis and Maguire, "Deconstructing Episodic Memory with Construction"; Carruthers, *The Craft of Thought.*

8. THE FEELING OF REMEMBERING

133 *deep in a pine forest* This fictional memory is adapted from the description in Demis Hassabis, Dharshan Kumaran and Eleanor A. Maguire, "Using Imagination to Understand the Neural Basis of Episodic Memory," *Journal of Neuroscience*, vol. 27, 2007, pp. 14365–74, Supplementary Table S1.

135 *technique known as EEG* Martin A. Conway, Christopher W. Pleydell-Pearce, Sharron E. Whitecross and Helen Sharpe, "Neurophysiological Correlates of Memory for Experienced and Imagined Events," *Neuropsychologia*, vol. 41, 2003, pp. 334–40.

137 *imagination inflation* Maryanne Garry, Charles G. Manning, Elizabeth F. Loftus and Steven J. Sherman, "Imagination Inflation: Imagining a Childhood Event Inflates Confidence That It Occurred," *Psychonomic Bulletin & Review*, vol. 3, 1996, pp. 208–14.

137 *proposing to a Pepsi machine* John G. Seamon, Morgan M. Philbin and Liza G. Harrison, "Do You Remember Proposing Marriage to the Pepsi Machine? False Recollections from a Campus Walk," *Psychonomic Bulletin & Review,* vol. 13, 2006, pp. 752–6.

137 *skin sample from a little finger* Giuliana Mazzoni and Amina Memon, "Imagination Can Create False Autobiographical Memories," *Psychological Science*, vol. 14, 2003, pp. 186–8.

137 *Imagination inflation . . . in future-oriented thinking* Karl K. Szpunar and Daniel L. Schacter, "Get Real: Effects of Repeated Simulation and Emotion on the Perceived Plausibility of Future Experiences," *Journal of Experimental Psychology: General*, vol. 142, 2013, pp. 323–7.

138 *fictional new popcorn product* Priyali Rajagopal and Nicole Votolato Montgomery, "I Imagine, I Experience, I Like: The False Experience Effect," *Journal of Consumer Research*, vol. 38, 2011, pp. 578–94.

139 *source monitoring framework* **Marcia K. Johnson, "Memory and Reality,"** *American Psychologist*, 2006, pp. 760–71; Karen J. Mitchell and Marcia K. Johnson, "Source Monitoring 15 Years Later: What Have We Learned from fMRI about the Neural Mechanisms of Source Memory?," *Psychological Bulletin*, vol. 135, 2009, pp. 638–77.

140 *false memories produced differing patterns* Craig E. L. Stark, Yoko Okado and Elizabeth F. Loftus, "Imaging the Reconstruction of True and False

Memories Using Sensory Reactivation and the Misinformation Paradigms," *Learning & Memory*, vol. 17, 2010, pp. 485–8.

140 *anterior medial prefrontal cortex* Jon S. Simons, "Constraints on Cognitive Theory from Neuroimaging Studies of Source Memory," in F. Rösler, C. Ranganath, B. Röder and R. H. Kluwe (eds.), *Neuroimaging of Human Memory: Linking Cognitive Processes to Neural Systems*, Oxford, Oxford University Press, 2009, pp. 405–26; Jon S. Simons, Richard N. A. Henson, Sam J. Gilbert and Paul C. Fletcher, "Separable Forms of Reality Monitoring Supported by Anterior Prefrontal Cortex," *Journal of Cognitive Neuroscience*, vol. 20, 2008, pp. 447–57.

142 *the "Memory Wars"* Daniel L. Schacter, *Searching for Memory: The Brain, the Mind, and the Past*, New York, Basic Books, 1996, chapter 9.

142 *memories of childhood sexual abuse* Elke Geraerts, Jonathan W. Schooler, Harald Merckelbach, Marko Jelicic, Beatrijs J. A. Hauer and Zara Ambadar, "The Reality of Recovered Memories: Corroborating Continuous and Discontinuous Memories of Childhood Sexual Abuse," *Psychological Science*, vol. 18, 2007, pp. 564–8; **Richard J. McNally and Elke Geraerts, "A New Solution to the Recovered Memory Debate,"** *Perspectives on Psychological Science*, vol. 4, 2009, pp. 126–34.

143 *false memory formation* Elke Geraerts, D. Stephen Lindsay, Harald Merckelbach, Marko Jelicic, Linsey Raymaekers, Michelle M. Arnold and Jonathan W. Schooler, "Cognitive Mechanisms Underlying Recovered-memory Experiences of Childhood Sexual Abuse," *Psychological Science*, vol. 20, 2009, pp. 92–8.

143 *"The difference between false memories and true ones . . ."* Salvador Dalí, *The Secret Life of Salvador Dalí* (translated by Haakon M. Chevalier), New York, Dover, 1993 (original work published 1942), p. 38.

143 *"Memory and the Law"* The British Psychological Society Research Board, *Guidelines on Memory and the Law: Recommendations from the Scientific Study of Human Memory*, Leicester, British Psychological Society, June 2008.

143 *studies . . . to determine whether reported memories are true or false* Daniel M. Bernstein and Elizabeth F. Loftus, "How to Tell If a Particular Memory Is True or False," *Perspectives on Psychological Science*, vol. 4, 2009, pp. 370–74; Aldert Vrij, "Criteria-based Content Analysis: A Qualitative Review of the First 37 Studies," *Psychology, Public Policy, and Law*, vol. 11, 2005, pp. 3–41; Martin A. Conway, Alan F. Collins, Susan E. Gathercole

and Stephen J. Anderson, "Recollections of True and False Autobiographical Memories," *Journal of Experimental Psychology: General*, vol. 125, pp. 69–95; Elisa Krackow, "Narratives Distinguish Experienced from Imagined Childhood Events," *American Journal of Psychology*, vol. 123, 2010, pp. 71–80.

144 *the "this is real" tag* Note that the source-monitoring model does not require the "tagging" of mental states as real or imaginary; rather, distinguishing between real and imaginary representations is a process of inference based on many different forms of information.

144 *memory repackaged as a premonition* Anthony P. Morrison, "The Use of Imagery in Cognitive Therapy for Psychosis: A Case Example," *Memory*, vol. 12, 2004, pp. 517–24; Mitchell and Johnson, "Source Monitoring 15 Years Later."

145 *the paracingulate sulcus* Marie Buda, Alex Fornito, Zara M. Bergström and Jon S. Simons, "A Specific Brain Structural Basis for Individual Differences in Reality Monitoring," *Journal of Neuroscience*, vol. 31, 2011, pp. 14308–13. Note that the participants scanned in this study were all healthy, apparently cognitively intact members of the general population.

9. REMEMBER ME A STORY

148 *"locked vault"* Catherine Loveday and Martin A. Conway, "Using Sense-Cam with an Amnesic Patient: Accessing Inaccessible Everyday Memories," *Memory*, vol. 19, 2011, pp. 697–704.

148 *Henry Molaison* Benedict Carey, "H.M., an Unforgettable Amnesiac, Dies at 82," *The New York Times*, December 4, 2008; Brian Newhouse, "H.M.'s Brain and the History of Memory," *NPR*, February 24, 2007, http://www.npr.org/templates/story/story.php?storyId=7584970; Brenda Milner, "Amnesia Following Operation on the Temporal Lobes," in C. W. M. Whitty and O. L. Zangwill (eds.), *Amnesia*, London, Butterworths, 1966.

149 *SenseCam* The device is now marketed commercially as the Vicon Revue. See http://research.microsoft.com/en-us/um/cambridge/projects/sensecam/.

150 *"lifelogging"* Joshua Foer, *Moonwalking with Einstein: The Art and Science of Remembering Everything*, London, Allen Lane, 2011, chapter 7; Leo Benedictus, "How I Remember: The Lifelogger," *Guardian*, January 14, 2012.

150 *Mrs. B* Emma Berry et al., "The Use of a Wearable Camera, SenseCam, as a Pictorial Diary to Improve Autobiographical Memory in a Patient with Limbic Encephalitis: A Preliminary Report," *Neuropsychological Rehabilitation*, vol. 17, 2007, pp. 582–601.

151 *highly visual nature of autobiographical memory* Loveday and Conway, "Using SenseCam with an Amnesic Patient."

151 *how viewing SenseCam images relates to changing activations in the brain* Martin A. Conway and Catherine Loveday, "Accessing Autobiographical Memories," in John H. Mace (ed.), *The Act of Remembering: Toward an Understanding of How We Recall the Past*, Oxford, Wiley-Blackwell, 2010, pp. 56–70.

154 *their own lives look so strange* R. Harper, D. Randall, N. Smythe, C. Evans, L. Heledd and R. Moore, "The Past Is a Different Place: They Do Things Differently There," *Proceedings of DIS08 Design Interactive Systems*, 2008, pp. 271–80.

155 *Proustian moments* Loveday and Conway, "Using SenseCam with an Amnesic Patient," p. 697.

156 *autobiographical knowledge structures* Martin A. Conway, "Memory and the Self," *Journal of Memory and Language*, vol. 53, 2005, pp. 594–628.

156 *"cognitive feelings"* Martin A. Conway, "Autobiographical Memory and Consciousness," in William P. Banks (ed.), *Encyclopedia of Consciousness*, vol. 1, Oxford, Academic Press, 2009, pp. 77–82.

157 *eighty-year-old former engineer* Christopher J. A. Moulin, Martin A. Conway, Rebecca G. Thompson, Niamh James and Roy W. Jones, "Disordered Memory Awareness: Recollective Confabulation in Two Cases of Persistent Déjà Vécu," *Neuropsychologia*, vol. 43, 2005, p. 1364.

157 *"routine memory glitch"* Alan S. Brown, "Getting to Grips with Déjà Vu," *The Psychologist*, vol. 17, 2004, p. 694; Alan S. Brown, "A Review of the Déjà Vu Experience," *Psychological Bulletin*, vol. 129, 2003, pp. 394–413.

158 *recollective experience* Moulin et al., "Disordered Memory Awareness."

159 *theta oscillation* Akira R. O'Connor, Colin Lever and Chris J. A. Moulin, "Novel Insights into False Recollection: A Model of Déjà Vécu," *Cognitive Neuropsychiatry*, vol. 15, 2010, pp. 118–44.

161 *confabulation is a perfectly understandable response* Gianfranco Dalla Barba and Caroline Decaix, "Do You Remember What You Did on March 13, 1985? A Case Study of Confabulatory Hypermnesia," *Cortex*, vol. 45,

2009, pp. 566–74; Paul W. Burgess and Tim Shallice, "Confabulation and the Control of Recollection," *Memory*, vol. 4, 1996, pp. 359–411; Armin Schnider, "Spontaneous Confabulation and the Adaptation of Thought to Ongoing Reality," *Nature Reviews Neuroscience*, vol. 4, 2003, pp. 662–71; Asaf Gilboa, "Strategic Retrieval, Confabulations, and Delusions: Theory and Data," *Cognitive Neuropsychiatry*, vol. 15, 2010, pp. 145–80.

162 *confabulations serve the needs of the self* Schacter, *How the Mind Forgets and Remembers*; Aikaterini Fotopoulou, "The Affective Neuropsychology of Confabulation and Delusion," *Cognitive Neuropsychiatry*, vol. 15, 2010, pp. 38–63.

166 *feeling overwhelmed and tearful* Bonnie-Kate Dewar and Fergus Gracey, "Am Not Was: Cognitive-behavioural Therapy for Adjustment and Identity Change Following Herpes Simplex Encephalitis," *Neuropsychological Rehabilitation*, vol. 17, 2007, pp. 602–20.

167 *The SenseCam gives Claire a way out of this impasse* Claire has said that because SenseCam pictures aren't "staged" like typical photographs are, they show her experiences as they really are, with the images giving a "much wider vision of what has actually happened and lots of fine detail about the reality of it."

171 *Memory . . . is also a way of being with other people* Berry et al., "The Use of a Wearable Camera, SenseCam, as a Pictorial Diary."

10. THE HORROR RETURNING

176 *Emotion does strange things to memory* Alexandre Schaefer and Pierre Philippot, "Selective Effects of Emotion on the Phenomenal Characteristics of Autobiographical Memories," *Memory*, vol. 13, 2005, pp. 148–60.

176 *the assassination of President Lincoln* F. W. Colegrove, "Individual Memories," *American Journal of Psychology*, vol. 10, 1899, pp. 228–55. One respondent recalled: "My father and I were on the road to A—— in the State of Maine to purchase the 'fixings' needed for my graduation. When we were driving down a steep hill into the city we felt that something was wrong. Everybody looked so sad, and there was such terrible excitement that my father stopped his horse, and leaning from the carriage called: 'What is it, my friends? What has happened?' 'Haven't you heard' was the reply—'Lincoln has been assassinated.' The lines fell from my father's limp hands, and with tears streaming from his eyes he sat as one bereft of

motion. We were far from home, and much must be done, so he rallied after a time, and we finished our work as well as our heavy hearts would allow" (pp. 247–8).

176 *flashbulb memories* Roger Brown and James Kulik, "Flashbulb Memories," *Cognition*, vol. 5, 1977, pp. 73–99.

177 *the resignation of the prime minister Margaret Thatcher* Martin A. Conway et al., "The Formation of Flashbulb Memories," *Memory & Cognition*, vol. 22, 1994, pp. 326–43.

178 *Challenger space shuttle disaster* Ulric Neisser and Nicole Harsch, "Phantom Flashbulbs: False Recollections of Hearing the News about *Challenger*," in E. Winograd and U. Neisser (eds.), *Affect and Accuracy in Recall: Studies of "Flashbulb" Memories*, Cambridge, Cambridge University Press, 1992.

178 *the terrorist attacks of 9/11* William Hirst et al., "Long-term Memory for the Terrorist Attack of September 11: Flashbulb Memories, Event Memories, and the Factors That Influence Their Retention," *Journal of Experimental Psychology: General*, vol. 138, 2009, pp. 161–76.

178 "... *Confidence often goes hand in hand with accuracy* ..." Ingfei Chen, "How Accurate Are Memories of 9/11?," *Scientific American*, September 6, 2011.

179 *events from our personal family history* David C. Rubin and Marc Kozin, "Vivid Memories," *Cognition*, vol. 16, 1984, pp. 81–95.

179 *activation leads to the release of hormones and neurotransmitters* Rachel Yehuda, Marian Joëls and Richard G. M. Morris, "The Memory Paradox," *Nature Reviews Neuroscience*, vol. 11, 2010, pp. 837–9.

181 *"War not only kills and wounds* ..." N. Greenberg, E. Jones, N. Jones, N. T. Fear and S. Wessely, "The Injured Mind in the UK Armed Forces," *Philosophical Transactions of the Royal Society B*, vol. 366, 2011, p. 261.

181 *lifetime incidence of the disorder* Rachel Yehuda, "Post-traumatic Stress Disorder," *New England Journal of Medicine*, vol. 346, 2002, pp. 108–14; Naomi Breslau and Ronald C. Kessler, "The Stressor Criterion in DSM-IV Post-traumatic Stress Disorder: An Empirical Investigation," *Biological Psychiatry*, vol. 50, 2001, pp. 699–704.

181 *"trauma" is extremely hard to pin down* Richard J. McNally, *Remembering Trauma*, Cambridge, MA, Harvard University Press, 2003, chapter 3.

182 *"haunted by the past"* Yehuda et al., "The Memory Paradox," p. 837.

182   *uncontrollability of memories* McNally, *Remembering Trauma*; Yehuda, "Post-traumatic Stress Disorder."

182   "... *Colorado, not Vietnam...*" Daniel L. Schacter, *Searching for Memory: The Brain, the Mind, and the Past*, New York, Basic Books, 1996, p. 210.

183   *involuntary memories* Dorthe Berntsen, "Involuntary Memories of Emotional Events: Do Memories of Traumas and Extremely Happy Events Differ?," *Applied Cognitive Psychology*, vol. 15, 2001, pp. 135–58.

183   *telephone survey* Dorthe Berntsen and David C. Rubin, "The Reappearance Hypothesis Revisited: Recurrent Involuntary Memories after Traumatic Events and in Everyday Life," *Memory & Cognition*, vol. 36, 2008, pp. 449–60, Study 2.

184   *diary of their involuntary memories* Ibid., Study 3.

184   *study of Vietnam veterans* McNally, *Remembering Trauma*, p. 115.

185   *filter of our later emotional states* Ingfei Chen, "A Feeling for the Past," *Scientific American*, January–February 2012, pp. 24–31.

185   *baffled gynecologist* M. I. Good, "The Reconstruction of Early Childhood Trauma: Fantasy, Reality, and Verification," *Journal of the American Psychoanalytic Association*, vol. 42, 1994, pp. 79–101.

185   *"weapon focusing"* Linda J. Levine and Robin S. Edelstein, "Emotion and Memory Narrowing: A Review and Goal-relevance Approach," *Cognition and Emotion*, vol. 23, 2009, pp. 833–75; McNally, *Remembering Trauma*, chapter 2.

185   *more general failings of memory* McNally, *Remembering Trauma*, chapter 5; Joseph LeDoux, *The Emotional Brain: The Mysterious Underpinnings of Emotional Life*, London, Weidenfeld & Nicolson, 1998.

186   *suppress other memories* Michael C. Anderson and Collin Green, "Suppressing Unwanted Memories by Executive Control," *Nature*, vol. 410, 2001, pp. 366–9; Michael C. Anderson et al., "Neural Systems Underlying the Suppression of Unwanted Memories," *Science*, vol. 303, 2004, pp. 232–5; Michael C. Anderson and Benjamin J. Levy, "Suppressing Unwanted Memories," *Current Directions in Psychological Science*, vol. 18, 2009, pp. 189–94; Ingrid Wickelgren, "Trying to Forget," *Scientific American*, January–February 2012, pp. 33–9.

186   *picture of a naked person* Stephen R. Schmidt, "Outstanding Memories: The Positive and Negative Effects of Nudes on Memory," *Journal of*

Experimental Psychology: Learning, Memory, and Cognition, vol. 28, 2002, pp. 353–61.

187 *smaller hippocampi* Rachel Yehuda and Joseph LeDoux, "Response Variation Following Trauma: A Translational Neuroscience Approach to Understanding PTSD," *Neuron*, vol. 56, 2007, pp. 19–32; Mark W. Gilbertson et al., "Smaller Hippocampal Volume Predicts Pathologic Vulnerability to Psychological Trauma," *Nature Reviews Neuroscience*, vol. 5, 2002, pp. 1242–7.

187 *surviving a car crash* LeDoux, *The Emotional Brain*, pp. 200–204.

188 *no good scientific evidence* Richard J. McNally and Elke Geraerts, "A New Solution to the Recovered Memory Debate," *Perspectives on Psychological Science*, vol. 4, 2009, pp. 126–34; Schacter, *Searching for Memory*; McNally, *Remembering Trauma*.

189 *death camps* Bruno Bettelheim, "Individual and Mass Behavior in Extreme Situations," *Journal of Abnormal and Social Psychology*, vol. 38, 1943, pp. 417–52; Bruno Bettelheim, *The Informed Heart: Autonomy in a Mass Age*, Glencoe, IL, Free Press, 1960, p. 155; Douwe Draaisma, *Why Life Speeds Up as You Get Older: How Memory Shapes Our Past* (translated by Arnold and Erica Pomerans), Cambridge, Cambridge University Press, 2004, pp. 118–19.

190 *repetition enhances memory* Schacter, *Searching for Memory*, chapter 9; McNally, *Remembering Trauma*, chapter 6.

190 *dissociation* McNally, *Remembering Trauma*, chapter 6.

193 *the "forgot-it-all-along" effect* Jonathan W. Schooler, Miriam Bendiksen and Zara Ambadar, "Taking the Middle Line: Can We Accommodate Both Fabricated and Recovered Memories of Sexual Abuse?," in M. A. Conway (ed.), *Recovered Memories and False Memories*, Oxford, Oxford University Press, 1997; Martin A. Conway, "Memory and the Self," *Journal of Memory and Language*, vol. 53, pp. 594–628; McNally and Geraerts, "A New Solution to the Recovered Memory Debate."

193 *"free-floating" fragments* Arthur P. Shimamura, "Commentary on Clinical and Experimental Approaches to Understanding Repression," in J. Don Read and D. Stephen Lindsay (eds.), *Recollections of Trauma: Scientific Evidence and Clinical Practice*, New York, Plenum Press, 1997.

193 *"knit together the relevant fragments . . ."* Schacter, *Searching for Memory*, p. 110.

194 *hippocampal system is not yet mature* W. J. Jacobs and Lynn Nadel, "Stress-induced Recovery of Fears and Phobias," *Psychological Review*, vol. 92, 1985, pp. 512–31.

194 *"behavioral memories"* Lenore C. Terr, "Forbidden Games: Post-traumatic Child's Play," *Journal of the American Academy of Child Psychiatry*, vol. 20, 1981, pp. 741–60; Lenore C. Terr, "What Happens to Early Memories of Trauma? A Study of Twenty Children Under Age Five at the Time of Documented Trauma Events," *Journal of the American Academy of Child and Adolescent Psychiatry*, vol. 27, 1988, pp. 96–104; McNally, *Remembering Trauma*, pp. 117–18.

195 *documented trauma such as sexual abuse* Ingrid M. Cordón, Margaret-Ellen Pipe, Liat Sayfan, Annika Melinder and Gail S. Goodman, "Memory for Traumatic Experiences in Early Childhood," *Developmental Review*, vol. 24, 2004, pp. 101–32; Terr, "What Happens to Early Memories of Trauma?"; Lenore Terr, *Unchained Memories: True Stories of Traumatic Memories, Lost and Found*, New York, Basic Books, 1994.

198 *new, emerging interpretation* Martin A. Conway, Kevin Meares and Sally Standart, "Images and Goals," *Memory*, vol. 12, 2004, pp. 525–31.

200 *the power of EMDR* Andrew Parker and Neil Dagnall, "Effects of Bilateral Eye Movements on Gist Based False Recognition in the DRM Paradigm," *Brain and Cognition*, vol. 63, 2007, pp. 221–5; Andrew Parker and Neil Dagnall, "Effects of Handedness and Saccadic Bilateral Eye Movements on Components of Autobiographical Recollection," *Brain and Cognition*, vol. 73, 2010, pp. 93–101; **Christian Jarrett, *The Rough Guide to Psychology*,** London, Rough Guides, 2011, chapter 25; Scott O. Lilienfeld, "EMDR Treatment: Less Than Meets the Eye?," http://www.quackwatch .org/01QuackeryRelatedTopics/emdr.html. EMDR is currently recommended as a treatment for PTSD by the UK's National Institute for Clinical Excellence (NICE), "Post-traumatic Stress Disorder (PTSD): The Treatment of PTSD in Adults and Children, NICE Clinical Guideline 26"; see http://www.nice.org.uk/nicemedia/pdf/CG026publicinfo.pdf.

201 *the traumatized mind* Shona Illingworth, *The Watch Man/Balkaniel*, London, Film and Video Umbrella, 2011; Wickelgren, "Trying to Forget."

201 *"Forgetting is no solution . . ."* Yehuda et al., "The Memory Paradox," p. 839.

11. THE MARTHA TAPES

207 *ready to look back* Daniel L. Schacter, *Searching for Memory: The Brain, the Mind, and the Past*, New York, Basic Books, 1996, chapter 10.

210 "*. . . you can close your eyes and summon it at will . . .*" Sarah Crown, "A Life in Books: Penelope Lively," *Guardian*, July 25, 2009.

210 *the reminiscence bump seems to be a basic fact* Douwe Draaisma, *Why Life Speeds Up as You Get Older: How Memory Shapes Our Past* (translated by Arnold and Erica Pomerans), Cambridge, Cambridge University Press, 2004, chapter 13; Martin A. Conway, "Memory and the Self," *Journal of Memory and Language*, vol. 53, pp. 594–628; Ashok Jansari and Alan J. Parkin, "Things That Go Bump in Your Life: Explaining the Reminiscence Bump in Autobiographical Memory," *Psychology and Aging*, vol. 11, 1996, pp. 85–91; Annette Bohn and Dorthe Berntsen, "The Reminiscence Bump Reconsidered: Children's Prospective Life Stories Show a Bump in Young Adulthood," *Psychological Science*, vol. 22, 2011, pp. 197–202.

211 *time seems to go past more quickly* Draaisma, *Why Life Speeds Up*, chapter 14; William James, *The Principles of Psychology*, vol. 1, New York, Cosimo, 1890, p. 625; Robert Krulwich, "Why Does Time Fly By as You Get Older?," NPR, http://www.npr.org/templates/story/story.php?storyId=122322542; Joshua Foer, *Moonwalking with Einstein: The Art and Science of Remembering Everything*, London, Allen Lane, 2011, p. 77. For findings that counter the life-speeding effect, see William J. Friedman and Steve M. J. Janssen, "Aging and the Speed of Time," *Acta Psychologica*, vol. 134, 2010, pp. 130–41.

212 *deficits in time perception* Teresa McCormack, Gordon D. A. Brown, Elizabeth A. Maylor, Lucy B. N. Richardson and Richard J. Darby, "Effects of Aging on Absolute Identification of Duration," *Psychology and Aging*, vol. 17, 2002, pp. 363–78.

213 *memory is like "a great plain . . ."* Hilary Mantel, *Giving Up the Ghost*, London, HarperCollins, 2003, p. 19; Hilary Mantel, "Father Figured," *Daily Telegraph*, April 23, 2005.

214 *older adults find it hard to keep track* Trey Hedden and John D. E. Gabrieli, "Insights into the Ageing Mind: A View from Cognitive Neuroscience," *Nature Reviews Neuroscience*, vol. 5, 2004, pp. 87–96; Catriona D. Good, Ingrid S. Johnsrude, John Ashburner, Richard N. A. Henson, Karl J. Friston and Richard S. J. Frackowiak, "A Voxel-based Morphometric Study of

Ageing in 465 Normal Adult Human Brains," *NeuroImage*, vol. 14, 2001, pp. 21–36.

214 *Ronald Reagan* Schacter, *Searching for Memory*, p. 287.

214 *how much information about source* Jon S. Simons, Chad S. Dodson, Deborah Bell and Daniel L. Schacter, "Specific- and Partial-source Memory: Effects of Aging," *Psychology and Aging*, vol. 19, 2004, pp. 689–94.

215 *incorrigible gossips* Schacter, *Searching for Memory*, p. 288.

216 *generally less vivid* Ibid.; Shahin Hashtroudi, Marcia K. Johnson and Linda D. Chrosniak, "Aging and Qualitative Characteristics of Memories for Perceived and Imagined Complex Events," *Psychology and Aging*, vol. 5, 1990, pp. 119–26.

216 *semantic memory is preserved* Brian Levine, Eva Svoboda, Janine F. Hay, Gordon Winocur and Morris Moscovitch, "Aging and Autobiographical Memory: Dissociating Episodic from Semantic Retrieval," *Psychology and Aging*, vol. 17, 2002, pp. 677–89; Jon Simons, "Can Memory Improve with Age?," *Guardian*, January 14, 2012.

217 *imagining future occurrences* Donna Rose Addis, Alana T. Wong, and Daniel L. Schacter, "Age-related Changes in the Episodic Simulation of Future Events," *Psychological Science*, vol. 19, 2008, pp. 33–41.

218 *the resignation of Margaret Thatcher* Gillian Cohen, Martin A. Conway and Elizabeth A. Maylor, "Flashbulb Memories in Older Adults," *Psychology and Aging*, vol. 9, 1994, pp. 454–63.

218 *memories of the invasion* Dorthe Berntsen and Dorthe K. Thomsen, "Personal Memories for Remote Historical Events: Accuracy and Clarity of Flashbulb Memories Related to World War II," *Journal of Experimental Psychology: General*, vol. 134, 2005, pp. 242–57.

221 *the emotional texture of an event* Johnson, "Memory and Reality"; Hashtroudi et al., "Aging and Qualitative Characteristics of Memories for Perceived and Imagined Complex Events"; Hedden and Gabrieli, "Insights into the Ageing Mind."

222 *narrative provides a helpful structure* Levine et al., "Aging and Autobiographical Memory"; Schacter, *Searching for Memory*, chapter 10. We do of course rely on stock memories throughout our lives, such as in our flashbulb memories and accounts of significant events such as giving birth. In old age, our weakened ability to remember recent events means that we become progressively more dependent on these relatively consistent narrative memories.

223 *"But what if, even at a late stage . . ."* Julian Barnes, *The Sense of an Ending*, London, Jonathan Cape, 2011, p. 120.

224 *crucial contextual factor is language* Viorica Marian and Ulric Neisser, "Language-dependent Recall of Autobiographical Memories," *Journal of Experimental Psychology: General*, vol. 129, 2000, pp. 361–8; Steen Folke Larsen, Robert W. Schrauf, Pia Fromholt and David C. Rubin, "Inner Speech and Bilingual Autobiographical Memory: A Polish-Danish Cross-cultural Study," *Memory*, vol. 10, 2002, pp. 45–54; Akiko Matsumoto and Claudia J. Stanny, "Language-dependent Access to Autobiographical Memory in Japanese-English Bilinguals and US Monolinguals," *Memory*, vol. 14, 2006, pp. 378–90.

225 *language reversion* K. D. Bot and M. Clyne, "Language Reversion Revisited," *Studies in Second Language Acquisition*, vol. 11, 1989, pp. 167–77; Monika S. Schmid and Merel Keijzer, "First Language Attrition and Reversion among Older Migrants," *International Journal of the Sociology of Language*, vol. 200, 2009, pp. 83–101.

227 *Kovno* Martha's ability to recognize the name of her mother's hometown is consistent with findings that recognition memory is unimpaired in the elderly; see Levine et al., "Aging and Autobiographical Memory." This description of the Yiddish interview formed the basis of Charles Fernyhough, "Unlocking Martha's Stories," *Guardian*, March 26, 2011.

232 *correcting for her own distortions of time* Susan E. Crawley and Linda Pring, "When Did Mrs. Thatcher Resign? The Effects of Ageing on the Dating of Public Events," *Memory*, vol. 8, 2000, pp. 111–21.

233 *"willful forgetting"* Rebecca Solnit, *A Field Guide to Getting Lost*, Edinburgh, Canongate, 2006, p. 47.

12. A SPECIAL KIND OF TRUTH

237 *processes of protein synthesis* See note for p. 13 *long-term potentiation*.

237 *factors such as sleep* Bjoern Rasch and Jan Born, "Maintaining Memories by Reactivation," *Current Opinion in Neurobiology*, vol. 17, 2007, pp. 698–703; Robert Stickgold, "Memory in Sleep and Dreams: The Construction of Meaning," in S. Nalbantian, Paul M. Matthews and James L. McClelland (eds.), *The Memory Process: Neuroscientific and Humanistic Perspectives*, Cambridge, MA, MIT Press, 2011, pp. 73–95; Jessica D. Payne, Robert Stickgold, Kelley Swanberg and Elizabeth A. Kensinger, "Sleep

Preferentially Enhances Memory for Emotional Components of Scenes," *Psychological Science*, vol. 19, 2008, pp. 781–8.

238 *". . . only as good as your last memory . . ."* Joseph LeDoux, article at http://www.edge.org/q2008/q08_1.html; Jonah Lehrer, *Proust Was a Neuroscientist*, Edinburgh, Canongate, 2011, chapter 4.

238 *A reconsolidating brain might not do much reconstruction* For further discussion, see Oliver Hardt, Einar Örn Einarsson and Karim Nader, "A Bridge over Troubled Water: Reconsolidation as a Link between Cognitive and Neuroscientific Memory Research Traditions," *Annual Review of Psychology*, vol. 61, 2010, pp. 141–67.

238 *scene construction experiment* Demis Hassabis, Dharshan Kumaran, Seralynne D. Vann and Eleanor A. Maguire, "Patients with Hippocampal Amnesia Cannot Imagine New Experiences," *PNAS*, vol. 104, 2007, pp. 1726–31; Sinéad L. Mullally, Demis Hassabis and Eleanor A. Maguire, "Scene Construction in Amnesia: An fMRI Study," *Journal of Neuroscience*, vol. 32, 2012, pp. 5646–5653.

239 *narrative is a key organizational force* **David C. Rubin, "The Basic-systems Model of Episodic Memory,"** *Perspectives on Psychological Science*, vol. 1, 2006, pp. 277–311; Raymond A. Mar, "The Neural Bases of Social Cognition and Story Comprehension," *Annual Review of Psychology*, vol. 62, 2011, pp. 103–34. For details of Bartlett's study of how people reshape stories to fit their own knowledge structures, see p. 12.

240 *Excuse me just a moment . . .* Hilary Mantel, *Wolf Hall*, London, Fourth Estate, 2009, p. 414.

241 *a kind of active remembering* Keith Oatley, "Review of *The Emigrants, The Rings of Saturn*, and *Vertigo* by W. G. Sebald," *Literary Review of Canada*, vol. 8, 2000, pp. 19–21.

241 *fictional memory-making* Charles Fernyhough, "Slippery Memories and the Tasks of Fiction," OnFiction, December 28, 2010, http://www.onfiction.ca/2010/12/slippery-memories-and-tasks-of-fiction.html; Charles Fernyhough, "The Story of the Self," *Guardian*, January 14, 2012.

242 *"Though my early memories are patchy . . ."* Hilary Mantel, *Giving Up the Ghost*, London, HarperCollins, 2003, p. 18. See note for p. 9 *Memoir is an increasingly popular literary genre.*

243 *One elderly refugee* Robert Fisk, *Pity the Nation: Lebanon at War*, Oxford, Oxford University Press, 1990, chapter 2.

243 *As a society, we are "remembering"* Charles B. Stone, Alin Coman, Adam D. Brown, Jonathan Koppel and William Hirst, "Toward a Science of Silence: The Consequences of Leaving a Memory Unsaid," *Perspectives on Psychological Science*, vol. 7, 2012, pp. 39–53; Tuğçe Kurtiş, Glenn Adams and Michael Yellow Bird, "Generosity or Genocide? Identity Implications of Silence in American Thanksgiving Commemorations," *Memory*, vol. 18, 2010, pp. 208–24; William Hirst et al., "Long-term Memory for the Terrorist Attack of September 11: Flashbulb Memories, Event Memories, and the Factors That Influence Their Retention," *Journal of Experimental Psychology: General*, vol. 138, 2009, pp. 161–76; Marianne Hirsch, "The Generation of Postmemory," *Poetics Today*, vol. 29, 2008, pp. 103–28.

244 *implications for our legal systems* BPS Research Board, *Memory and the Law*; Elizabeth Loftus, "Our Changeable Memories: Legal and Practical Implications," *Nature Reviews Neuroscience*, vol. 4, 2003, pp. 231–4; Benjamin Weiser, "In New Jersey, Rules Are Changed on Witness IDs," *The New York Times*, August 24, 2011; Laura Beil, "The Certainty of Memory Has Its Day in Court," *The New York Times*, November 28, 2011.

244 *(hypothetical) drug to erase the memory of a trauma* Eryn J. Newman, Shari R. Berkowitz, Kally J. Nelson, Maryanne Garry and Elizabeth F. Loftus, "Attitudes about Memory Dampening Drugs Depend on Context and Country," *Applied Cognitive Psychology*, vol. 25, 2011, pp. 675–81.

244 *"memory-dampening" procedures* Larry Cahill, Bruce Prins, Michael Weber and James L. McGaugh, "β-Adrenergic Activation and Memory for Emotional Events," *Nature*, vol. 371, 1994, pp. 702–4; Diancai Cai, Kaycey Pearce, Shanping Chen and David L. Glanzman, "Protein Kinase M Maintains Long-term Sensitization and Long-term Facilitation in *Aplysia*," *Journal of Neuroscience*, vol. 31, 2011, pp. 6421–31; Daniela Schiller, Marie-H. Monfils, Candace M. Raio, David C. Johnson, Joseph E. LeDoux and Elizabeth A. Phelps, "Preventing the Return of Fear in Humans Using Reconsolidation Update Mechanisms," *Nature*, vol. 463, 2010, pp. 49–53; Katie Drummond, "No Fear: Memory Adjustment Pills Get Pentagon Push," *Wired*, December 16, 2011; Adam Kolber, "Neuroethics: Give Memory-altering Drugs a Chance," *Nature*, vol. 476, 2011, pp. 275–6; Adam Piore, "Totaling Recall," *Scientific American Mind*, January–February 2012, pp. 40–45; Charles Fernyhough, "Total Recall," *BBC Focus*, August 2012, pp. 54–59.

245 *deep-brain stimulation* Adrian W. Laxton et al., "A Phase I Trial of Deep Brain Stimulation of Memory Circuits in Alzheimer's Disease," *Annals of Neurology*, vol. 68, 2010, pp. 521–34.

245 *reliance on Google* Betsy Sparrow, Jenny Liu and Daniel M. Wegner, "Google Effects on Memory: Cognitive Consequences of Having Information at Our Fingertips," *Science*, vol. 333, 2011, pp. 776–8.

246 *"own special kind" of truth* Salman Rushdie, *Midnight's Children*, New York, Avon, 1980, p. 253.

246 *If our memories are constructions* Fernyhough, "The Story of the Self."

246 *"If life has a base that it stands upon . . ."* Virginia Woolf, "A Sketch of the Past," in *Moments of Being* (2nd edition), Orlando, FL, Harcourt Brace & Company, 1985, p. 64.

248 *". . . our memories must see double . . ."* Roger Shattuck, *Proust's Binoculars: A Study of Memory, Time, and Recognition in "A La Recherche du Temps Perdu,"* Princeton, Princeton University Press, 1983, p. 47.

ACKNOWLEDGMENTS

I have relied on the expertise of many people in writing *Pieces of Light*. The idea for the book developed when I was a Fellow of the Institute of Advanced Study at Durham University. Since that time, a great number of people have provided advice, information and useful discussion: Terri Apter, Alan Baddeley, David Bainbridge, Helen Beer, Sima Beeri, Clare Connor, Stephen Connor, Martin Conway, Mo Costandi, Barry Davis, Christine Dyer, Alex Easton, Fadia Faqir, Justine Fieth, Katerina Fotopoulou, Maria Georgiou, Rhett Griffiths, Omar Robert Hamilton, Demis Hassabis, Markus Hausmann, Hazel Hirshorn, Sadie Hirshorn, Shona Illingworth, Simon James, Keith Laws, Catherine Lloyd, Elizabeth Loftus, Tim Lott, Robert Macfarlane, Sara Maitland, Andrew Mayes, Teresa McCormack, Anthony McGregor, Kevin Meares, Chris Moulin, Paula Nolan, Seamus Nolan, Keith Oatley, Silvia Orr, Richard Osbourne, Derek Pay, Edward Platt, Catherine Robson, Corinne Saunders, Sophie Scott, Toby Scott, Raja Shehadeh, Marc Smith, Sid Smith, Sally Standart, Maren Stange, Tom Stoddart, David Sutton, Jon Sutton, Kim Wade, Marina Warner, Pat Waugh, Susanne Weis and Rebekah Willett. I owe a particularly large debt to those who shared stories that, for reasons of privacy, remain anonymous.

Several people read and commented on versions of the manuscript:

Christian Jarrett, Catherine Loveday, Elizabeth Meins, Catriona Morrison, Dan Schacter, Hugo Spiers, Valerie Webb and Angela Woods. Jon Simons not only read large chunks of the book but has also been an incomparable source of wisdom and helpful debate. Needless to say, all errors and omissions are my own.

My agent, David Grossman, has been a constant source of support throughout the book's development. At Profile, my editor Lisa Owens provided careful and insightful readings and tremendously constructive editorial input. Sarah Caro saw the potential in the book early on and provided valuable suggestions in the early stages. Daniel Crewe, Penny Daniel, Andrew Franklin, Rebecca Gray and their colleagues at Profile contributed expertise at critical points, and Sally Holloway was an exemplary copyeditor. At HarperCollins, Gail Winston, Maya Ziv, Heather Drucker and their colleagues made the preparation of the US edition a huge pleasure. I am also grateful to Isabel Berwick, Nell Card, Harriet Green, Claire Malcolm, Natalie Roper, Julia Wardropper and my many friends and colleagues at the School of Life.

My greatest debt is to my family—Lizzie, Athena and Isaac—whose love, patience and support have made this book possible.

INDEX

Page numbers in *italics* refer to illustrations.

ABOUT THE AUTHOR

Charles Fernyhough is an award-winning writer and psychologist. His previous book, *A Thousand Days of Wonder: A Scientist's Chronicle of His Daughter's Developing Mind*, was a *Parade* magazine pick of the week and has been translated into seven languages. The author of two novels, *The Auctioneer* and *A Box of Birds*, Fernyhough has written for the *Guardian*, the *Financial Times* and the *Sunday Telegraph*; contributes to public radio's *Radiolab*; blogs for *Psychology Today*; and is a Reader in Psychology at Durham University, UK. *Pieces of Light* was a *Sunday Times*, *Sunday Express* and *New Scientist* book of the year. Visit his website at www.charlesfernyhough.com.